CALIFORNIA AMPHIBIANS

AND HOW TO FIND THEM

Foothill Yellow-legged Frog. *Photograph by Marisa Ishimatsu.*

CALIFORNIA AMPHIBIANS

AND HOW TO FIND THEM

EMILY TAYLOR

H

HEYDAY

Berkeley, California

Library of Congress Cataloging-in-Publication Data

Names: Taylor, Emily (Lecturer in biological sciences) author
Title: California amphibians and how to find them / Emily Taylor.
Description: Berkeley, California : Heyday, [2026] | Includes
  bibliographical references.
Identifiers: LCCN 2025019548 | ISBN 9781597146999 paperback | ISBN
  9781597147002 epub
Subjects: LCSH: Salamanders--California | Frogs--California |
  Salamanders--California--Identification |
  Frogs--California--Identification
Classification: LCC QL653.C2 T394 2026 | DDC 597.809794--dc23/
eng/20250903
LC record available at https://lccn.loc.gov/2025019548

Cover Art: Spencer Riffle
Cover Design: Marlon Rigel, based on the original design by Debbie Berne
Interior Design/Typesetting: theBookDesigners, based on the original design
by Debbie Berne

Published by Heyday
P.O. Box 9145, Berkeley, California 94709
(510) 549-3564
heydaybooks.com

Printed in East Peoria, Illinois, by Versa Press, Inc.

10 9 8 7 6 5 4 3 2 1

To my students,
past, present, and future.
You are the main reasons I have gone to work
with a smile every day for twenty years
and will for twenty more.

Coastal Giant Salamander. *Photograph by Spencer Riffle.*

Western Spadefoot. *Photograph by Ryan Sikola.*

CONTENTS

Lesser (left) and Black-bellied (right)
Slender Salamanders.
Photograph by Spencer Riffle.

Family Salamandridae

THE FROGS

Family Ascaphidae

Family Bufonidae

Family Eleutherodactylidae

Family Hylidae

Pacific Chorus Frog. *Photograph by Spencer Riffle.*

California Tiger Salamander. *Photograph by Marisa Ishimatsu.*

PREFACE

What comes to your mind when I mention the word "frog"? Perhaps a small, leggy animal, green and cute, with big eyes. Maybe a frog from pop culture hopped into your mind. The empathetic and optimistic Kermit the Frog was the best friend of numerous American children starting in the 1950s. In the 1980s, the pioneering video game *Frogger* hooked millions of us kids on helping the little green creatures cross the road. The contemporary Pokémon universe features many frog-like characters, including Froakie, Toxicroak, and Bulbasaur.

While frogs feature heavily in pop culture and therefore the minds of people, salamanders are almost entirely absent. California's amphibians—the frogs and the salamanders—are like yin and yang in this way and others. Frogs are conspicuous because they are loud, communicating with a language like we do. In the winter and spring, we hear choruses of hopeful little suitors hollering, "Mate with me! No, not him—me!" at discerning females. In contrast, salamanders are silent. They are more secretive than frogs, and most species of salamander are rarely seen unless you're looking for them. Few Californians know a thing about the salamanders around them, and many don't know much more about the frogs.

Despite their relative anonymity, amphibians enjoy a fairly good reputation. Ceramic sculptures of fat toads and jumping frogs adorn gardens. People even buy little toad houses to put in their yards. Amphibians aren't venomous (though some are poisonous, a distinction I explain in the book), so they are pretty much harmless to people. Frogs feature prominently in many Indigenous tales, but their standing is a bit different. In tales from both the Pomo people and the Chumash people, droughts are blamed on Frog, who stalwartly refuses to allow the trickster Coyote to drink from the last remaining

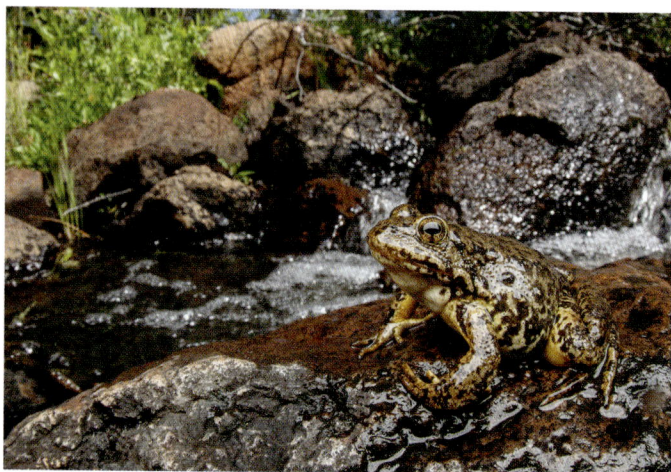

Sierra Nevada Yellow-legged Frog.

sources of water. Many of the lands where I go out searching for amphibians are the ancestral homelands of the Pomo and Chumash people, and I wrote this book on my small ranch on Salinan ancestral territory. It is my honor to help promote respect for this land and for the people who have resided and currently reside here by sharing my knowledge about its amphibian wildlife.

Amphibians are a group of animals that bring me boundless joy, and it is my goal in this book to share some of it with you. It's a good thing that I don't have to be stingy with said joy because I have enough of it to go around! California's wealth of salamander life ranges from the tiny, terrestrial slender salamanders with heads that don't seem big enough to hold brains to the aptly named giant salamanders that can be as long as your forearm and perfectly embody the *amphibious* (traversing water and land) lifestyle. On the other

Northwestern Salamander.

side of the California amphibian tree lies a retinue of equally fascinating creatures, with highlights including warty and charismatic toads and treefrogs that bely their names by climbing around on decidedly non-tree surfaces. You'll learn all the details in the species accounts, which are designed to wow you for the purpose of wooing you into love for all things amphibian. These accounts are served up with a side dish of how to find them in the wild because while viewing animals at zoos is totally cool, searching for and finding amphibians in the wild is an unparalleled experience.

I begin the introduction of this book by discussing the importance of water for amphibians, then explore what exactly we mean when we use the words "frog" or "salamander" while introducing readers to the major groups of amphibians that occupy California. The meat of this book consists of accounts describing the natural history of each species and detailed information on how to find them in the wild. I chose to combine certain species into single accounts when their natural history is very similar. This book is not a field guide complete with range maps and detailed distinguishing characteristics of each species; rather, my goal is to introduce Californians and visitors to our fascinating cast of amphibian characters. I also discuss how (and when) to properly capture and handle amphibians, with numerous cautionary notes about the dangers of unwittingly harming them.

Which brings me to the final, and most important, theme that arises again and again in this book: amphibian declines. Amphibians in California and the rest of the world face grave threats from climate change, habitat destruction, pollution, disease, and introduction of exotic species. No other single group of animals is more imperiled.

Let's take the California Tiger Salamander as an example. This is a California endemic species, meaning that it occurs here in California and nowhere else. It once occupied much of coastal California and the Central Valley, but populations have been decimated by habitat destruction, isolation and inbreeding, genetic hybridization with an introduced species that pollutes its bloodlines, and multiple diseases, such that the California Tiger Salamander could go extinct in our lifetime without the heroic efforts of conservation biologists and land managers.

I have a recurring fantasy of traveling to California's Central Valley in a time machine back to the pre–Gold Rush era, where I would be surrounded by vast wetlands instead of farms and oil rigs. I can just picture the millions of waterfowl alternately landing and taking off from the enormous lake that spread across the valley; the river tributaries so thick with salmon that you could practically walk across on their backs; the woodlands rustling with stealthy wolves, Grizzly Bears, and jaguars. In the fall, millions of California

Tiger Salamanders would have migrated overland from mammal burrows to ponds in a giant mass of squirming black-and-yellow limbs and tails. The waterbirds would have feasted on their larvae like seagulls attacking abandoned french fries, but millions of salamanders would still have made it to metamorphose into healthy adults to start the cycle all over again.

In the past 150 years, the Central Valley has been almost completely converted to farmland and oil fields to sate the never-ending hunger of California's human population and its vehicles. Most large predators like Grizzly Bears and jaguars are long gone, and California endemic amphibians and reptiles like the California Tiger Salamander, the Blunt-nosed Leopard Lizard, and the Giant Garter Snake are absent from the vast majority of their previous ranges. Sometimes I can feel the loss of such biomass and biodiversity as physical pain. This is made worse by the fact that few Californians know, let alone care, about what California used to be before its forced metamorphosis into its modern state, and what we still stand to lose if we continue on the way we have been.

In this book, I describe the threats to California amphibians and suggest ways that we can help them. Even though you are just one of 40 million residents and visitors that occupy California at any given time, our collective actions amount to a tangible effort that can supplement the incredible conservation programs designed by biologists, zoos, and land managers that are saving amphibian lives en masse. For you, it starts with this book: read it, learn how to find the amphibians in your area, go out and observe these incredible animals in the wild, and repeat. Share this book and your newfound skills with others, especially children. Because real, live amphibians in the wild beat Kermit, Frogger, and Pokémon any day.

Coastal Tailed Frog tadpole. *Photograph by Spencer Riffle.*

INTRODUCTION

California: Amphibians' Wild Water Park

As a child, I absolutely adored water parks. Summer visits to my cousins in the inland Los Angeles area meant much-anticipated trips to Raging Waters, where we whooshed down water slides over and over until our pink skin needed to be soothed with aloe vera and our bellies needed to be filled with In-N-Out. Now, as an adult, I leave the slides to the young folks and park myself on an inner tube on the lazy river at my local water park in Paso Robles, where the combination of warm breeze, cool water, and fun people-watching can't be beat.

These water parks bear some parallels with California's amphibian-filled wetlands. The kiddie pool is crowded with little tadpoles gathered near its edges. In the deeper pools, young males

Pacific Chorus Frogs are one of the most common amphibian species in California.

splash around, hollering at females and trying to one-up each other in playful games of aquatic courtship. Adults sit in the shade alongside the water, jumping in occasionally when their skin needs to be cooled, then hoisting themselves back out to dry off or waddle away to find a snack.

The word *amphibian* means "part water, part land," and the name celebrates the incredibly important moment in our evolutionary history when an ancient, fish-like ancestor used its limbs to propel itself from the water onto land. This step was not instantaneous but took millions of years of evolution. The eventual result of the conquering of land by early amphibians is a shocking diversity of modern-day amphibians that rely on water more than most other animals. California is a veritable water park for our amphibians, providing a multitude of wetlands where frogs and salamanders mate, get food, cool down, escape from predators, and so much more.

There are no truly marine amphibians, so it is fresh water that we are talking about here. California's vast landscape is slashed with scores of amphibian-laced rivers, streams, and creeks, and stippled with thousands of lakes, ponds, and meadows wet with snowmelt. Though not ideal habitat, our vast farmlands, irrigation canals, cattle impoundments, and roadside ditches provide watery havens for some hardy frog species. Even on my little ranch on the arid central coast, toads and chorus frogs have colonized the tiny, year-round wetlands I unwittingly created beneath my hot tub and inside my irrigation shed.

But it is not just the collective size of California's water bodies, historic and current, that make it a wild water park for amphibians. Compared to many other states, California boasts a huge diversity of habitat types, climates, and plant life that weave complex tapestries of interconnected life. The rainy forests of extreme Northern California house sensitive and remote stream habitats,

Two Western Toads bask in a wetland. *Photograph by Zeev Nitzan Ginsburg.*

where humans seldom barge in on the local frogs and salamanders. Mountain lakes in the Sierra Nevada were sanctuaries for the now endangered yellow-legged frogs that are starting to make a come-back after careful protection of their habitat. Even the vast deserts of southeastern California boast amphibian species that thrive around their streams, oases, and seeps.

California's waters are not all pristine, however. They, too, are threatened by destruction, drying out, and pollution from the hordes of hungry and thirsty people that call our state home. Amphibians are especially sensitive to pollution and other environmental disturbances. Their skin is permeable to facilitate gas and water exchange, which unfortunately also forms a weak barrier to pollutants, and many amphibians have two life stages, one in water and one on land, making them susceptible to stressors in both habitats.

Southern Mountain Yellow-legged Frogs rely on clean water for all aspects of their lives. *Photograph by Max Roberts.*

For these reasons and more, many amphibian species in California are in trouble, which I describe more fully in this introduction and in each relevant species account.

Before we dive more deeply into the lives of California's amphibians as well as ways we can support them, let's step back to discuss what an amphibian actually is and learn some important guiding principles and vocabulary relevant to the world of amphibians that you will meet in this book.

What Are Amphibians?

The Amphibia is a group that consists of highly water-dependent animals that you might casually know as frogs and salamanders. However, there is a difference between the colloquial words we use in everyday English and the technical terms scientists use, and here I introduce you to those terms as they apply to California amphibians. Beyond frogs and salamanders, there is a whole other major group of amphibians called caecilians (pronounced like "Sicilians"), which is a mysterious lineage of limbless amphibians that live in fresh water or in soil in tropical areas. There are no caecilians remotely close to California, so we will not discuss them further here and instead turn our attention to the familiar four-legged amphibians we can find lurking, loping, hopping, and swimming around the Golden State.

This book organizes California amphibians by group, starting with orders Caudata (salamanders) and Anura (frogs and their relatives). Salamanders are introduced first because they are evolutionarily a more *basal* group than frogs, meaning that they diverged earlier from the vertebrate evolutionary tree. Within those orders, all amphibians are presented alphabetically by their family (name ends in -*idae*, like Ranidae) then by their scientific name, consisting of italicized genus and species (like *Rana aurora*). Throughout the book, I refer to species mainly by their common names (e.g., Northern Red-legged Frog).

Salamanders are tailed amphibians in the group Caudata

The first order of amphibians in California is the salamanders, the tailed amphibians that go by the technical name Caudata ("tailed"). In California these caudatans all have four legs, giving them a lizard-like shape, though elsewhere on the globe there are examples of species with no hind limbs. Of the more than eight hundred species of described salamanders from ten families in the world, members of five of these families can be found in California.

The most *speciose* (with the highest number of species) salamander family in California is also the most speciose in the US and in the world: the Plethodontidae, or the "lungless salamanders." These salamanders are exceptions to all other California amphibians in that they lay eggs on land instead of in the water. They are still dependent on water, though, as their skin must be kept moist to facilitate breathing through it, given that they don't have lungs for this job. Both the salamanders and their eggs typically can be found in moist areas underground to prevent desiccation. Another family, Rhyacotritonidae, or "torrent salamanders," are related to plethodontids but live and lay eggs in the water. The other three families of California salamanders are truly amphibious, living part of the year on land and heading to ponds to mate and lay eggs during the

Spencer Riffle

The California Slender Salamander is a tiny and very common amphibian in parts of Northern California.

rainy season. These are the Ambystomatidae ("tiger salamanders" or "mole salamanders"), Dicamptodontidae ("giant salamanders"), and Salamandridae ("newts").

The group Anura consists of frogs and their relatives

The second group of amphibians we will discuss is the Anura ("no tail"), an order consisting of the tailless amphibians that we often collectively call frogs. While the technical term for all these animals is anurans, we can be forgiven for referring to them as frogs for the sake of simplicity. This vast group is extremely diverse, with fifty-seven families and more than 7,700 species so far described by biologists worldwide. In California, there are just five families of native anurans plus species from two introduced families. Let's explore these families to get the hang of the terminology we use to refer to these amphibians.

California Chorus Frog.

The Ranidae are known as "true frogs" and include the classic long-legged good swimmers and jumpers that you might picture when you think of a frog. Another family, Ascaphidae, is known as the "tailed frogs," though their "tail" is not actually a tail at all but something far more fascinating (you will need to read on to find out what). The family Hylidae is the "treefrogs," a group of anurans with enlarged toe pads that help them climb not just trees but practically any surface. The family Bufonidae consists of "true toads," the warty and adorably grumpy creatures folks tend to be familiar with. The last native family of California anurans is the Scaphiopodidae, or the "spadefoots," a fascinating group of anurans that spend most of the year underground in a barely animate state as they wait for seasonal rainfall to stimulate them to emerge and start a wild party of feeding and breeding, only to dig back underground when the fun is over. Members of two families have

Spade-like structures on the hind limbs of the aptly named spadefoots allow these anurans to burrow into the soil.

also been introduced to parts of California. These consist of the Common Coquí in the family Eleutherodactylidae and the African Clawed Frog in the family Pipidae, both strange aliens that you will learn about in their respective species accounts.

From amplexus to toadlets: more vocabulary you should know

In this book, I try to avoid technical terms when possible, but a short vocabulary lesson is in order so that you understand several key terms when you encounter them in the species accounts. The first series of terms involves reproduction and development. Prepare yourself to be amazed at what I am about to tell you. All amphibians have sex, but only one species of amphibian in California has sex in the way we people talk about it—via copulation. In all other amphibians, males don't have penises and sex typically occurs in one of two ways: 1) in salamanders, a male deposits

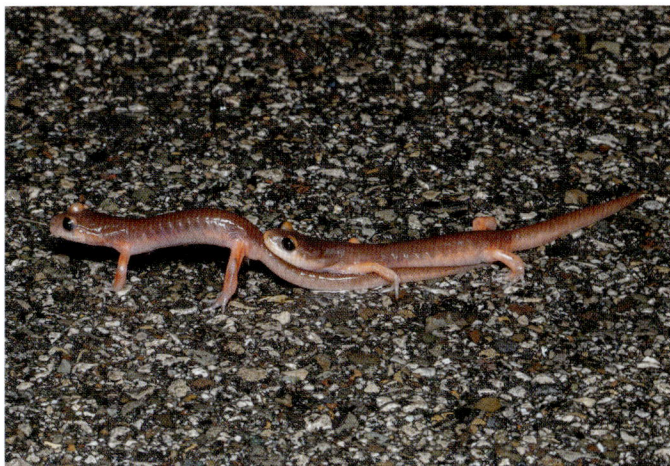

A male Ensatina (left) guides a female (right) in a courtship dance.

a sperm packet on the substrate (on land in plethodontids, and underwater in all others), and the female squats over it to pick it up with her cloaca, a pouch inside the sole "hole" of amphibians used for peeing, pooping, mating, and egg laying; or 2) in most frogs, a male grasps a female from behind in a position technically called *amplexus* and nontechnically (e.g., by me, when trying to be cute) called "froggy-style," then she lays her eggs in the water as he ejaculates onto them.

Development in terrestrial plethodontid salamanders is fairly simple. The embryos develop inside small, gelatinous eggs some-where moist to prevent them from drying out, like the rocky crevice shown in the photos on the next page. In many species, the mother salamander stays with the eggs and guards them from predators. The eggs then hatch into tiny "salamanderlings," which is not a technical term but should be, in my humble opinion.

Amorous Western Toads in a mating posture called amplexus. *Photograph by Max Roberts.*

This Wandering Salamander laid her eggs in this crevice and stayed with them until they hatched into salamanderlings. *Photographs by Spencer Riffle.*

All other California amphibians lay their eggs in the water. These eggs hatch into an aquatic juvenile life stage called *larvae* (singular *larva*). In anurans, the larvae have gill slits (leading to internal gills where breathing occurs) and no legs. It is common to call these larvae "tadpoles" or "pollywogs." After a period of feeding and development, the tadpole undergoes metamorphosis, where it changes into an adult form. On the outside, metamorphosis consists of tadpoles losing gills, sprouting limbs, and absorbing their tails, but many less visible changes occur on the inside, too. They develop reproductive organs, marking attainment of maturity. Their guts and mouthparts remodel as they transition from grazing tadpoles to carnivorous adults. Newly metamorphosed anurans are typically called *metamorphs* (also "froglets" or "toadlets," when appropriate).

In aquatic salamanders, the larvae have tufted external gills and four legs, and are not called tadpoles. Like frogs, they also lose their gills and develop reproductive organs when they metamorphose and move onto land. Fascinatingly, some members of the truly amphibious salamander families (Ambystomatidae, Dicamptodontidae, and

Larval Coastal Giant Salamander.

Salamandridae) metamorphose into reproductively active adults but retain their gills and stay in the water. This is called *neoteny* or *paedomorphosis*, both of which mean that the adult salamander retains juvenile characteristics. You might be familiar with Axolotls, an ambystomatid, paedomorphic (and excessively adorable) salamander native to Mexico and popular in the pet trade and pop culture.

Identifying California amphibians

Identifying amphibians to species requires examining several characteristics. First, since you cannot always determine size based on photos, I have described each species as either small (typically less than half the length of your finger), medium (about the size of the palm of your hand), or large (the length of your hand and fingers, or bigger). Next, the appearance of eggs and larvae often allow identification to species. In tadpoles, their mouthparts are especially useful for this purpose. In this book, I describe the general appearance of the eggs and larvae of each species but do not provide photos or show the mouthparts for all of them. I mainly focus on describing the appearance of adults. Remember—this is not a technical field guide but rather a basic reference geared to help you find, watch, learn about, and fall in love with the amphibians of California. Those of you seeking detailed information, including drawings of the mouthparts of California tadpoles, should check out the comprehensive field guide *California Amphibians and Reptiles*, by Robert Hansen and Jackson Shedd.

Mouthparts can be used to identify this Coastal Tailed Frog tadpole.
Photograph by Spencer Riffle.

Some California Amphibians Are in Trouble

As you read this book, you might feel a bit of whiplash. One second I may tell you a story about an amazing experience I had watching a salamander in the wild, and the next moment I may lament the disappearance of a species of toad from most (or, in some cases, all) of the state. The lighthearted goal of this book is to teach you how to find and observe California's amphibians in the wild so that you can enjoy them as I have had the pleasure of doing, but there is also a heavyhearted stepsister to that goal creeping around in the background. This is my need to spread the word about the dire straits we currently find many amphibian species in, often at our hands.

Scientists have known for at least a generation that all is not well with our beloved wildlife. A famous line in Aldo Leopold's 1949 book *A Sand County Almanac* reads, "One of the penalties of an

Spencer Riffle

Many species of amphibians, like this California Red-legged Frog, are threatened with extinction.

ecological education is that one lives alone in a world of wounds." He means that we, as scientists, are privy to the horrible reality of the threats facing wildlife while many members of the public remain blissfully unaware. Trust me, I would much rather fill this book with stories of sunny meadows bursting with smiling salamanders and frolicking frogs, but I would by lying.

It's depressing for me to tell you the truth, and it is undoubtedly sad for you to read about it. But the vast majority of people don't know about modern amphibian declines, or that California is one of the top five hotspots on the planet for threatened amphibian species. Half of the seventy or so species of native amphibians that currently inhabit California are protected by the state or federal governments because their populations are very small or declining, in some cases precipitously. When a species is thought to be at risk, the California Department of Fish and Wildlife designates it as a species of special concern. If the situation becomes more dire, either the state or the federal government (or both) can elevate a species' status to threatened or even endangered, if the species is at serious risk of extinction.

It's hard to know when exactly the trouble started, but it was almost certainly when development of California's land began en masse in the early and middle parts of the twentieth century. Worldwide, 40 percent of all amphibian species are threatened (that's over 3,500 species!), and hundreds of species have gone extinct in the past 150 years due to human activities, including four species in California: Sonoran Desert Toad, Arizona Toad, Oregon Spotted Frog, and Lowland Leopard Frog (all four species live on in neighboring states). Like the California Tiger Salamander I describe in the preface to this book, California amphibians have been deeply impacted by a combination of factors, including habitat destruction for agriculture, oil, and urbanization; pollution; water diversion for

Spencer Riffle

Some amphibians, like this Arboreal Salamander, do fairly well in habitats transformed by people. However, most amphibians do not.

farming; introduced species that eat or outcompete them; pathogens and parasites; climate change; and more. Though destruction of habitat is undoubtedly the leading cause of amphibian declines, another headlining villain is a fungal pathogen called chytrid (short for the fungus's scientific name, *Batrachochytrium dendrobatidis*), which has decimated amphibian populations around the world, including in California.

Why should we care about the plight of California amphibians? Here are a few notable reasons. Amphibians play extremely important roles in California's food webs, acting as major predators of invertebrates and as prey to other wildlife. As amphibious animals, they help with cycling nutrients back and forth between aquatic and terrestrial habitats. Amphibian declines can upset the natural balance in our ecosystems, impacting far more than just the frogs and salamanders themselves. A 2024 study identified amphibians as *the* most threatened type of animal in the world due to climate change caused by global emissions. Mark Urban, the lead author of the study, noted that the amphibian species in danger of extinction could hold biological secrets that we risk never learning. In a recent interview on NPR, he said, "Each of these species has encountered and solved life's problems, and so they're really the great books of knowledge on Earth. And we really don't want to burn those books before we get a chance to read them."

The importance of trying to help amphibians goes beyond the animals themselves. Coal miners used to take canaries in little cages with them into the depths of the mines. When the birds keeled over from exposure to toxic gases, it was time for the miners to return to the surface to avoid being similarly poisoned. The high sensitivity of birds to toxic gases has made the phrase "canary in a

Sierra Nevada Yellow-legged Frog tadpoles. Amphibians are particularly sensitive to pollutants because of their permeable skin and reliance on water. *Photograph by Max Roberts.*

coal mine" a metaphor for a harbinger of problems that will affect people if we ignore them and allow them to compound. Due to their permeable skin and reliance on water, amphibians are the quintessential canaries in the California coal mine, where their struggles with climate change, pollution, disease, and so many other threats are signals of what could happen to us if we don't take action. Amphibian biologist Tyrone Hayes studies how a common herbicide pollutant can harm developing frogs and warns people, "Your sons and daughters develop in water, just like my tadpoles do."

What Can We Do to Help California Amphibians?

So, what are we going to do about this mess? Actions to help California amphibians fall into two categories: scientists and managers working to save populations and species, and people making small changes to their lifestyles and engaging in activism that can collectively add up to make a real difference in helping protect our frogs and salamanders. In the species accounts in this book, you will read about the specific threats that face California's imperiled amphibian species and about the steps being taken by scientists, state and federal wildlife and land managers, and the public to help them. For some species, careful observation is all that is needed at this time. But in some of the most threatened species, it is not enough to just collect data—interventions by scientists are necessary to save them.

One example of such an intervention is rearing them in human care at zoos. Some people think that zoos are merely places where the public can view animals for their own amusement. But what you see when you visit a zoo is just a tiny part of what these incredible organizations do. Behind the scenes, many zoos have staff scientists and veterinarians that are working on sophisticated solutions to wildlife problems. Collaborative programs between scientists, land managers, and zoos are helping augment and repatriate frog populations in places where their numbers have dropped very low or where they have gone extinct due to habitat degradation, infections by the chytrid fungus, and other stressors.

For example, the San Francisco Zoo has a conservation center with staff who collect eggs or tadpoles of these species, rear them at the zoo, then release them back into the wild once they grow up. These "head-starting" programs work because zoo staff know better than anyone else how to rear animals, including how to treat them with antifungal solutions to kill the chytrid. In the wild, most eggs and tadpoles would not make it to adulthood due to predation and disease, so head-starting them bypasses that mortality and ends with the reintroduction of healthy, fat adults into the

Marisa Ishimatsu

These soon-to-be-released California Red-legged Frogs were head-started at a zoo to improve their chances of surviving and breeding.

wild, hopefully ready to mate and help the population grow natu-rally. And it is working! Joint efforts between the San Francisco Zoo, Oakland Zoo, National Park Service, US Forest Service, California Department of Fish and Wildlife, and UC Santa Barbara–affiliated Sierra Nevada Aquatic Research Laboratory have successfully repa-triated Sierra Nevada Yellow-legged Frogs (see page 198) and Cal-ifornia Red-legged Frogs (see page 175) in many areas of Yosemite National Park and other wetlands in California.

The San Diego Zoo Wildlife Alliance has taken on the daunting task of managing the Southern Mountain Yellow-legged Frog, which is in even more trouble than the Sierra Nevada Yellow-legged Frog and California Red-legged Frog. The Southern Mountain Yellow-legged Frog has a very limited distribution in mountains in Southern Califor-nia and the southern Sierra Nevada, and their populations have been decimated by chytrid fungus, drought, invasive fishes, and wildfires. They are very close to being extinct in the wild, even with the zoo program sustaining them by breeding and head-starting them, then releasing them as metamorphs and adults. But it doesn't stop there. Complex problems require serious research to solve them, and the San Diego Zoo Wildlife Alliance and their government partners are on the front lines. Current projects include testing whether frogs pro-duce more offspring if they are kept at low temperatures the winter before breeding, using artificial fertilization to improve the number of embryos produced, crossbreeding frogs from different mountain ranges to reduce inbreeding, and stimulating the immune systems of the frogs to make them better at resisting the chytrid fungus once released. Without these efforts, many more populations of California amphibians would be gone.

While these programs are amazing, it is not enough to leave all the work to the scientists. If you are like me, reading about the plight of our beloved amphibians is simultaneously frustrating and

A researcher releases a head-started Southern Mountain Yellow-legged Frog in the San Bernardino Mountains.

energizing. Only with knowledge about the state of our amphibians can you be expected to do anything about it, in the forms of voting, activism, food choices, energy usage, landscaping of your yard, and so much more. Truthfully, the problems are complex and there is a lot to be done. But there is also a long menu of things you can do to help, ranging from simple changes to complicated lifestyle overhauls.

What will YOU do to help amphibians? The next page is a list of suggestions to get you started.

What you can do to protect California amphibians

Vote—Prioritize representatives and propositions that protect water and open space.

Go green—Recycle; use less water, electricity, and plastic; avoid using pesticides and herbicides.

Go clean—Sanitize your hiking boots with a mild bleach solution between hikes to avoid spreading pathogens, especially if you visit wetlands.

Remove litter—Clean up trash, especially near wetlands, including plastic, glass, old fishing gear, etc.

Keep your cats indoors—Cats kill billions (yes, billions!) of wild animals every year, including amphibians. Dogs can also attack wildlife, so keep your eyes on them, too.

Rewild—Diversify your backyard with amphibian-friendly landscaping, including native plants, water sources, and cover objects . . . and leave the leaf litter alone.

Buy your plants locally, including Christmas trees—Avoid accidentally introducing exotic amphibians and other hitchhikers by planting locally grown native plants in your yard and cutting down your holiday tree at a local farm.

Spread the word—Share an amphibian story with a friend (or stranger!).

Foster amphibian love—Take your friends and family—especially kids—to see amphibians at the zoo and in the wild.

Donate—Support groups that protect amphibians and our land.

Get involved—Join a local herpetological society, club, newt brigade, or other amphibian-obsessed group. They would love to have you.

To Have and to Hold Amphibians

Most herpetologists I know tell stories of themselves as feral children running around their backyards or neighborhood parks, digging arthropods out of the dirt, catching lizards using a blade of grass as a lasso, and pouncing on slippery frogs. Kids are the ultimate hands-on learners, wanting to feel the texture of the skin of their prize and to see its little face and toes up close. Catching frogs is practically a rite of passage among outdoorsy kids.

I will never argue vehemently against holding an animal in your hands to see it up close, unless it's dangerous like a rattlesnake or prohibited due to protected status. There is something deeply valuable in the connection to nature that a child, or even an adult, can experience when communing with an animal up close and personal. That said, catching and holding animals, even briefly,

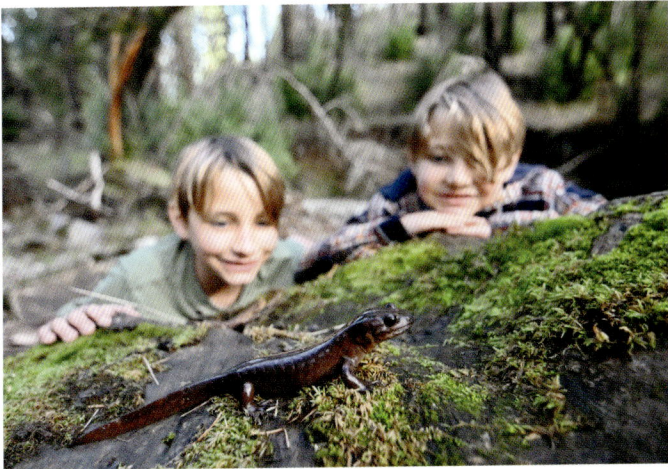

Two brothers admire a California Giant Salamander.

can sometimes be harmful to them, and teaching kids to watch amphibians without picking them up is a great idea.

In my first book, *California Snakes and How to Find Them*, I acknowledge that holding snakes can be a formative gateway experience into nature for some kids, but sometimes leaving the snakes alone allows us to witness incredible snake behaviors that we would have missed if we had grabbed them. In my second book, *California Lizards and How to Find Them*, I describe my own fascination with reptiles that was cemented in large part when I converted my dad's fishing pole into a lizard lasso, but I warned that mishandling lizards can cause them to drop their tails in a bid to escape. In this book, I will also take the balanced approach of singing the praises of holding an adorable California amphibian in your hands while warning of the potential stress that it could cause them. When learning how to find (and occasionally handle) amphibians, you should keep in mind these questions:

Will I hurt the amphibian by holding it?

Many of California's amphibians are tiny and fragile, which is why it's important that people do not roughly handle and injure them. For most salamanders, it is best to gently pick them up by the body and place them in the palm of your hand. Most frogs are wiggly, requiring you to hold them by the legs and waist firmly in your fist. Unlike the thick, protective scales of reptiles, the permeable nature of amphibian skin means they can readily absorb chemicals from our hands. If you have recently applied sunscreen, bug spray, or other chemicals to your skin, handling an amphibian can fatally poison them. Some amphibians carry fungi and other pathogens in their skin, which you could spread if you hold one individual right after another. Even clean hands can upset healthy microbial communities on amphibian skin. If you're going to hold an amphibian, be sure that you have washed your hands with soap and water

Max Roberts

The best way to handle amphibians is with gloves.

before doing so. Even better is to use nitrile gloves, which you can pick up at a drugstore.

Could the amphibian hurt me if I touch it?

There are no venomous amphibians, so you don't need to worry about dangerous bites or stings. However, several species of amphibians in California secrete toxins from their skin meant to deter predators, and these toxins could poison you if you handle one of these amphibians wrong. The typical distinction between a venom and a poison is that a venom is injected and a poison is ingested. In California, newts in the genus *Taricha* produce varying levels of a neurotoxin called tetrodotoxin, which, if ingested, can paralyze the breathing muscles of predators. Newts are not on the menu for people, and typically holding a newt is harmless, but you should avoid doing that if you have any scrapes or cuts on your hands. Some toads and spadefoots secrete noxious chemicals when handled that can irritate your eyes if you rub them after handling the creatures. This is yet another reason why using gloves is a great idea. No California amphibian species is remotely dangerous when handled with gloves.

Is handling this amphibian prohibited?

As you have now discovered, many amphibian species throughout the world, including in California, are imperiled due to factors like habitat loss, pollution, and disease. The result is that many California amphibians are protected by the state or federal governments. You can watch them, but you can't catch them, poke them to get a better angle for a photo, or otherwise interfere with them. At the end of each species account in this book, I include the protected status so that you can easily see which species are off limits for handling. Bear in mind that even common species are not to be handled in state and national parks, as well as in certain private preserves with their own rules.

If you really get into finding California amphibians in the wild, you will observe that the answer to one or more of these questions is "yes" for many species, and you therefore need to be content with watching them from a distance. Binoculars can help with watching some frog species as they sit on the bank of a pond or poke their heads out of the water. Most salamanders are harder to see without disturbing them by flipping the cover objects that they hide beneath. The last section of each species account answers the titular question of how to find them, but sometimes warns you away from doing this when the species is protected or when its habitat is sensitive. While this can be a hard pill to swallow, consider the fact that we humans have endangered our state's amphibians in the first place, so it is our responsibility to ensure that we do not disturb them further.

Despite the protected status of so many of California's amphibians, there are plenty of thriving species that you can go see in the wild in any corner of our state. With a fishing license from the California Department of Fish and Wildlife, you are permitted to catch, admire, and release all our unprotected amphibians. Take your family or friends out on a wet night and catch a Western Toad crossing the road to show them its warty skin. Join a neighborhood "newt patrol," a.k.a. "salamander bucket brigade," to help the creatures safely cross the roads at night during their winter migrations. Carefully turn over logs or flat rocks after the winter rains begin to look for the skin-breathing salamanders sheltering beneath. Go listen to the frogs calling for mates at night from ponds in your town. By learning about California amphibians and how to find them, you will unlock the door to the secret world of moist-skinned, four-legged fellows living in the trees, soil, and water around us.

In many neighborhoods, California residents form "newt patrols" to help the creatures cross busy roads at night. *Photograph by Doug Hofmockel.*

California Newts and their eggs.
Photograph by Spencer Riffle.

THE
SALAMANDERS

CALIFORNIA TIGER SALAMANDER

AMBYSTOMA CALIFORNIENSE

FAMILY AMBYSTOMATIDAE

It is fitting that the California Tiger Salamander is the first entry in this book! A California endemic species (found in this state and nowhere else), it is large, beautiful, and charismatic, just like California itself. This big salamander spends most of its time underground, but if you are lucky enough to see its boldly spotted form lumbering across the grass on its way to a watery tryst that will create the next generation, you will surely jump for joy.

Unfortunately, the California Tiger Salamander is not doing well. As a species with both aquatic and terrestrial life phases, they experience threats from both the water and the land. Habitat alteration has rendered over half of their historic geographic range uninhabitable. Grasslands and wetlands have been replaced by farms, oil fields, and housing developments. Many salamanders are run over by cars when trundling slowly to their mating ponds during the first fall rains. Water pollution has poisoned the habitats where they lay their eggs and their larvae develop. Invasive species like American Bullfrogs, Western Tiger Salamanders, and numerous fishes compete with California Tiger Salamanders for resources, eat them outright, or, in the case of the Western Tiger Salamander, hybridize with them and drown out the unique genes that define them as California Tiger Salamanders. These invasive species often carry diseases that impact this salamander and many other amphibians. Global warming and increased ultraviolet light exposure, due to thinning ozone in the atmosphere, may weaken the salamanders and exacerbate the other threats above. If populations become small enough, they may suffer from a lack of genetic diversity known as inbreeding depression and could blink out in a few generations without our intervention.

Luckily, the California Tiger Salamander enjoys both state and federal protection under the Endangered Species Act, which funnels millions of dollars annually to their management. Veritable armies of biologists in California are tasked with protecting the remaining populations of the California Tiger Salamander. Modern molecular techniques have allowed scientists to conduct scientifically informed crossbreeding of salamanders from different populations to recover the genetic "health" of inbred populations. Across their range, land managers have restored wetlands with California Tiger Salamanders in mind. Channeling the *Field of Dreams* mantra "If you build it, they will come," restored wetlands are often inundated with breeding California

Tiger Salamanders in a few short years. In just one story of many, the state of California harnessed the "wildness" of its vast military bases to protect sensitive species with a comprehensive California Tiger Salamander emergency protection plan at Travis Air Force Base in Northern California, which includes monitoring wetlands, diverting salamanders from crossing plane runways, and assisting them in crossing roads. This program has been a smashing success, with over three thousand California Tiger Salamanders observed on the base last year, less than 2 percent of which were dead or injured. Such actions, funded by your tax dollars, help ensure that the California Tiger Salamander and many other species will be there for years to come.

Appearance: These large salamanders are black with yellow or cream spots and have bulging eyes and big, round faces that appear to express perpetual smiles. The gelatinous eggs are laid singly or in rows attached to vegetation underwater. The aquatic larvae are a speckled gold-and-brown color and have three tufted gills on each side of the face, plus paddle-like tails.

Natural History: California Tiger Salamanders are highly reliant on the burrows of California Ground Squirrels and other burrowing rodents, and spend the majority of their time there. During fall and winter rains, these salamanders exit the burrows and amble overland to their breeding ponds, often the same pond in which they were born. A male courts a female by poking her with his snout, then he deposits a packet of sperm on the floor of the pond, which she picks up with her cloaca. She lays hundreds of tiny, yellowish eggs one at a time or in small clusters. The eggs hatch in a few weeks into small aquatic larvae that eat tiny aquatic invertebrates and other amphibian larvae, then they grow until the summer,

when they lose their gills and develop lungs as they metamorphose onto land. Adults eat insects and other invertebrates.

Range and Variations: California Tiger Salamanders once roamed grasslands and woodlands from the central coast through the San Joaquin and Sacramento River valleys to the foothills of the Sierra Nevada, but now they live only in isolated patches from Santa Barbara County north to Sonoma and Colusa Counties.

How to Find California Tiger Salamanders: There are two optimal ways to see California Tiger Salamanders. First, take a night drive along roads in California Tiger Salamander habitat during the first major storms of the fall. You can sometimes see them crossing the road as they disperse to their breeding ponds. Please be careful not to run over any salamanders, as they can be hard to see. Second, if you have access to land with their breeding ponds, you can look for them at night with a flashlight from late fall through the winter. At first you might see adults looking for mates, and later you might see the larvae resting near the bottoms of the water.

Protection: threatened (California), threatened or endangered (federal) depending on region

Marisa Ishimatsu

NORTHWESTERN SALAMANDER

AMBYSTOMA GRACILE

~~~~~~

FAMILY AMBYSTOMATIDAE

Northwestern Salamanders have one of the coolest adaptations in all of salamanderdom: green eggs. Winter rains bring mama salamanders with bloated abdomens out of their burrows to waddle overland to a pond, mate, and deposit their eggs in a huge mass attached to underwater vegetation. The Dr. Seussian eggs are green because they are infused with symbiotic algae that trade elements with the developing embryos inside the eggs. This is as necessary an arrangement as peanut butter and jelly; one just doesn't do very well without the other. The baseball-sized egg masses are so big and tightly packed that oxygen can't diffuse from the water into the embryos in the middle. No problem—the algae produce oxygen by photosynthesis while the embryos in the eggs hand over nitrogen in their waste to the algae. For over a hundred years, the algae were thought to move from pond water to infiltrate the salamander egg cells, and as a result the salamander algae's scientific name is *Oophila amblystomatis*, which charmingly translates to "he who loves salamander eggs." More recently, scientists have shown that the algae also are taken up by the embryos themselves. This means that the little developing salamanderlings can actually photosynthesize, making them the only photosynthetic vertebrates! It doesn't stop there—the embryos' immune systems somehow "know" to avoid attacking the friendly algal cells as foreign invaders, a cellular truce of sorts.

By the way, remember when I mentioned Christmas trees in the introduction? While editing the first draft of this book in December 2024, I saw a news article about a Northwestern Salamander that a Santa Barbara family recently found in their

Marisa Ishimatsu

Christmas tree. These salamanders are native to California, but they are (true to their name) a Pacific Northwest species that is found only in the far northwestern part of the state. Most Christmas trees sold in California are harvested in Oregon and Washington, right in the middle of the salamander's range, and sometimes Northwestern Salamanders take refuge under piles of recently cut trees and then end up accidentally hitchhiking to California. If the family had released the salamander in Santa Barbara, he could've potentially spread disease or "genetically polluted" native species if he could interbreed with them. Luckily, the family handed over the salamander (nicknamed Noël by the rescuers) to a wildlife rescue group, which then gave him to a local herpetologist who would use him as an educational animal. Upon further investigation, this has happened many times before, with Northwestern Salamanders riding along on Christmas trees as far as Florida! The moral of Noël's Christmas tale? If you celebrate Christmas, visit a local tree farm to cut down your tree. This will also prevent other hitchhikers, like insects, from spreading via the tree trade.

Let's take a look at the day-to-day lives of our "solarmanders" in the wild in northwestern California.

Appearance: Northwestern Salamanders are large, beefy salamanders that range in color from dark brown to black. They have conspicuous oval glands behind their eyes that give their heads a stocky look. The grooves along their sides are deep and prominent. Their eggs are laid in large, gelatinous masses, and they hatch into brownish-green larvae with three large gill filaments on each side.

Natural History: Northwestern Salamanders live in coastal grasslands and forests, where they spend much of the year hiding underground until they migrate to breeding wetlands

Symbiotic algae are evident in egg masses from a distance (top) and inside individual eggs (bottom). *Photographs by Spencer Riffle.*

during winter and spring rains. In the water, an amorous male grasps a female from behind like a spoon, vigorously giving her a head massage with his chin. He then drops a sperm packet on the floor of the pond or stream for her to pick up with her cloaca. She lays the enormous egg masses near the surface of the water, and after a few weeks of developing with the help of their algal friends, the eggs hatch into little aquatic larvae. These typically metamorphose into terrestrial, air-breathing adults later that year, but in some cases the salamanders become reproductively active yet remain in the aquatic, water-breathing form (paedo-morphosis). During aquatic phases, Northwestern Salamanders eat invertebrates and tadpoles, and adult salamanders on land eat terrestrial invertebrates.

Range and Variations: Northwestern Salamanders are found in only extreme northwestern California, from coastal Sonoma County northward through Oregon, Washington, and well into British Columbia.

How to Find Northwestern Salamanders: In California, most Northwestern Salamanders live in low-elevation coastal areas, where temperatures are mild enough that you can find them migrating to breeding ponds on rainy nights as early as January. If you can find their breeding ponds or streams, you might be able to watch them courting and mating. You can also flip logs near wetlands in the hopes of finding Northwestern Salamanders hiding underneath them.

Protection: none

LONG-TOED SALAMANDER

AMBYSTOMA MACRODACTYLUM

FAMILY AMBYSTOMATIDAE

When I decided to write this book after completing its predecessors on snakes and lizards of California, I made a list of the frog and salamander species of California that I had not yet seen in the wild. This was back in December 2022, and I admit to poring over it with even more excitement than my Christmas list that year. Why? Because it held the promise of a fun year of amphibian hunting to come. Long-toed Salamanders topped the list as a species that I wanted to observe in person, but they can be challenging to see. The southernmost subspecies, the Santa Cruz Long-toed Salamander, is perhaps the rarest salamander in California, even one of the rarest in the United States and the entire planet. One of their main breeding areas is a tiny area of wetland that is enclosed in a private ecological reserve. In 2015, severe drought completely dried out the breeding ponds, leaving the aquatic larvae gasping desperately in the dregs of the muddy sludge on the pond floors. Government scientists intervened and placed them inside rearing tanks, cared for them, and released them back into the ponds when conditions improved. Without this help, a large chunk of the last Santa Cruz Long-toed Salamanders on Earth could have blinked out in a single year.

On a wet night in January 2023, my graduate students and I donned our rain gear and headed out to a spot where it is possible see Santa Cruz Long-toed Salamanders if you are lucky: a tiny stretch of road outside the ecological reserve where they breed. We walked back and forth along this road, watching for salamanders in our flashlight beams. We found three of these salamanders that night, all females full of eggs and on their way to the ponds within the reserve, where hordes of male salamanders that arrived the previous month awaited their arrival. Seeing this salamander in the wild showed me one reason why it is so colorful: the yellow spots on its back help it blend in with the colors of the fallen oak and willow leaves in the area. I was pleased to see that the area is heavily monitored to protect this vulnerable little salamander but that people are still allowed to search for them along the roads. We were questioned once by a California Fish and Wildlife law enforcement agent, who approached us with a friendly "Y'all looking for salamanders?" and left us to continue our search. We were also stopped by a State Parks agent, but he was looking for people who had been involved in an altercation nearby. When he learned that this odd band of raincoat-and-headlamp-clad folks were "just out photographing salamanders," he laughed and said, "You are my kind of people," before speeding away. Take my advice, folks. Find a

group of kindred nerds to get into this exciting type of "trouble" on a Saturday night. Your life will be richer for it.

Appearance: Long-toed Salamanders are medium-sized salamanders, with dark background coloration punctuated by fat, yellowish-orange blotches down their backs and light speckling along their sides. Compared to most other salamanders that have webbed feet or short toes, these salamanders have long, thin toes, hence their name. The eggs are laid in the shallow water of ponds, sometimes attached to pieces of wood, singly or in small clusters. The aquatic larvae that hatch out have three long gill filaments on each side of their round faces, plus a fin running from their backs down their tails.

Natural History: Long-toed Salamander habitat is characterized by thick tree cover, often with fallen logs that provide moist habitat for much of the year, and they breed in shallow ponds. In some parts of their range, like the northern and central Sierra Nevada, they can be found at very high elevations. Long-toed Salamanders spend months hiding underground or inside wet logs, emerging during fall or winter rains to crawl overland and mate in ponds. Timing depends on climate; low-elevation populations usually mate early in the year, while high-elevation populations travel to breeding ponds as late as the spring. After mating, they crawl back onto land to find a spot to hide. Males typically hang out in the ponds longer than females, eating aquatic invertebrates while they wait for mates to arrive. Females eat mostly terrestrial invertebrates and spend less time in the ponds. Mating in Long-toed Salamanders can be as exciting as a mixed martial arts fight. Underwater, a male grasps a female around her armpits from behind, rubbing his head over her back. He moves forward and deposits a sperm packet for her to pick up with her cloaca. All the while, other males are trying to dislodge him from her back in a massive wrestling match. Some-times they even sneak in and throw down their own sperm packets in an attempt to trick her into picking up theirs instead! (These shenanigans sound exhausting—I can see why females spend as little time in the water as possible.) The females lay eggs in small, grape-like clusters or sometimes singly before leaving the pond. Larvae hatch out in a few weeks, feed on small aquatic invertebrates and tadpoles, and metamorphose into terrestrial adults in either a few months in populations with warmer climates or a few years in populations inhabiting colder areas. Sometimes Long-toed Sala-manders can become reproductively active without metamorphos-ing, keeping their gills and remaining in the water.

Range and Variations: Long-toed Salamanders have a patchy distribution in Northern California. There are two subspecies in the state: the Southern Long-toed Salamander (*Ambystoma mac-rodactylum sigillatum*) and the Santa Cruz Long-toed Salamander

(*Ambystoma macrodactylum croceum*). The Southern Long-toed Salamander is found in parts of Trinity, Siskiyou, Shasta, and Modoc Counties, plus it extends throughout the northern Sierra Nevada as far south as northern Tuolumne County. The Santa Cruz Long-toed Salamander is endemic to California, with a tiny distribution in Santa Cruz and Monterey Counties.

How to Find Long-toed Salamanders: Long-toed Salamanders are typically difficult to find outside of mating season. On wet late fall and winter nights in low-elevation sites, or on spring nights in colder sites at higher elevations, walk around with a flashlight to look for Long-toed Salamanders ambling toward their breeding ponds. If you can find their breeding ponds, you can shine your flashlight into the water to see the males waiting for females or grappling over them when they arrive. Long-toed Salamanders can also be found hiding under logs or other cover objects near the breeding ponds during the day.

Protection:
Southern Long-toed Salamander: species of special concern (California)
Santa Cruz Long-toed Salamander: endangered (California and federal)

Max Roberts

WESTERN TIGER SALAMANDER

AMBYSTOMA MAVORTIUM

FAMILY AMBYSTOMATIDAE

My very first job in herpetology was a summer gig after my junior year of college surveying the reptiles and amphibians of the Buffalo Gap National Grassland in South Dakota. One fun study that we conducted involved surveying prairie dog burrows for Western Tiger Salamanders. After sunset, we drove to one of the many clusters of burrows that are called "towns" in the southwestern part of the state and systematically peered inside the burrows with flashlights to look for salamanders. On occasion we would startle a Prairie Rattlesnake coiled inside, and one time

I found myself face-to-face with an irritated badger! However, over the course of our surveys, we found that the large nocturnal Western Tiger Salamanders peered out at us from a surprisingly high number of the burrows. We concluded that the prairie dog burrows were incredibly important as refugia for these salamanders and that the salamanders probably didn't help or hurt the resident prairie dogs in any way, making their relationship one of commensalism.

I may have fond feelings about the Western Tiger Salamander from that formative summer when I first became a herpetologist, but its story in California is not a happy one. Fishermen have long used the larvae of Western Tiger Salamanders as bait for predatory fishes, supporting a thriving industry of trade throughout the country. Fishermen sometimes dump unused bait into the water when they call it a day. As a result, Western Tiger Salamanders spread throughout the country and became established in many bodies of water where they weren't native. In many areas of California, the Western Tiger Salamanders hybridize with our native, endangered California Tiger Salamander, essentially watering down the genetic signature of the native species. Not only that, but the Western Tiger Salamander's monster-large larvae routinely eat the larvae of California Tiger Salamanders and other native amphibians. The escaped bait larvae also spread disease, including viruses and the deadly chytrid fungus that has decimated so many amphibian populations.

It sounds grim, doesn't it? I won't sugarcoat it. Invasive species like the Western Tiger Salamander are a huge threat to our native biodiversity in California. It is now illegal to sell this species as bait in California, but the damage may already be done. Research continues into how to mitigate the effects of this cute but dangerous non-native species in California.

A non-native Western Tiger Salamander in San Diego County. *Photograph by Jeff Lemm.*

Appearance: Western Tiger Salamanders are very large, second only in size to our native giant salamanders (genus *Dicamptodon*, see page 49). They have round faces and bulbous eyes, and are typically black or gray in color with cream or yellow bars or spots, though sometimes their colors and patterns can be more muted. The aquatic larvae are a speckled yellow-green, have three tufted external gills on each side of the head, and can grow as large as the adults.

Natural History: The natural history of the Western Tiger Salamander in California is similar to that of our native California Tiger Salamander (see page 30). They are found in many habitat types and breed in temporary or permanent wetlands, preferring still water, like ponds, over rapidly moving streams. Adults spend most of their time underground in the burrows of ground squirrels and other rodents, then migrate to wetlands when rains commence, where they mate and lay eggs that hatch in just over a week. The larvae eat invertebrates, tadpoles, salamander larvae, and small fishes, then metamorphose into the terrestrial adult form. Adults are so large that they eat just about anything. Unlike most populations of California Tiger Salamanders, in some cases the larvae of the Western Tiger Salamander become reproductively mature without metamorphosing.

Range and Variations: Western Tiger Salamanders have five subspecies recognized in their native range, and it is probable that multiple subspecies have been introduced to California. They have since hybridized with one another and with the California Tiger Salamander, making the variations in California a genetic mess. Numerous small populations of Western Tiger Salamanders can be found throughout California. If you are curious about known populations, you can consult iNaturalist to see their distribution in fine scale.

How to Find Western Tiger Salamanders: You can find Western Tiger Salamanders the way I did that summer doing surveys: by shining your flashlight inside rodent burrows at night. You can also drive around and find them crossing roads at night, especially during storms. Their large larvae are visible at the bottoms of wetlands, sometimes year-round.

Protection: none (non-native)

Coastal Giant Salamander. *Photograph by Spencer Riffle.*

GIANT SALAMANDERS
DICAMPTODON ENSATUS AND *D. TENEBROSUS*

FAMILY DICAMPTODONTIDAE

About ten years ago, I was invited to spend Thanksgiving weekend camping at a friend's property in the Santa Cruz Mountains. We each prepared a dish to share and enjoyed a chilly and damp yet beautiful weekend gathered around a campfire with friends. But the highlight of the trip appeared on my morning walk along a trail. I was ambling along, listening to an audiobook, when I nearly stumbled over a huge California Giant Salamander sitting right in front of me. The sheer size of these salamanders is impressive enough, but add in the fact that they are rare as far as salamanders go, and the result is that finding a giant salamander is a major reason to celebrate. The California Giant Salamander is endemic to California, found only in the vicinity of the San Francisco Bay Area, while a related species, the Coastal Giant Salamander, extends northward as far as Canada. I have combined them into a single account for this book because their biology is somewhat similar, at least based on the little we know about them, as rather few studies have been done on these charismatic amphibians.

What we do know about these secretive salamanders includes some biological zingers. For example, a population of California Giant Salamanders was recently discovered in an underground water-filled cave in Santa Cruz County. They never fully metamorphose, instead keeping their external gills and remaining in the water like larvae, but still developing their adult reproductive organs. In all other populations, this neoteny happens only occasionally, but here, metamorphosis into terrestrial forms wouldn't make sense since they all remain aquatic. This population has also lost its color pattern, with all individuals pale gray, which also makes sense because there is no selection for the camouflaging blotchy pattern in the dark environment of the cave. This population exemplifies the famous quote by evolutionary biologist Theodosius Dobzhansky that "Nothing in biology makes sense except in the light of evolution," though I would add a quip at the end, ". . . or in the dark of a cave."

Appearance: As their names imply, these salamanders are true giants. Their bodies, not including their tails, are as long as an adult person's hand. They have large heads with big eyes, and their tails are flattened sideways to help them swim when they mate. The background color is brown or brick red, with a copper-colored marbled pattern. Since the two species look similar, the best way to tell them apart is based on location, but the California Giant

Coastal Giant Salamander. *Photograph by Marisa Ishimatsu.*

Salamander (*Dicamptodon ensatus*) typically has markings on its pale chin whereas the Coastal Giant Salamander (*Dicamptodon tenebrosus*) does not. The eggs are seldom seen because they are hidden. Larvae are dark with fine, orangish mottling, tail fins, and small external gills.

Natural History: These salamanders inhabit wet forests in the vicinity of water, especially rocky streams. The adults are terrestrial, emerging from burrows or refugia under logs or rocks at night to hunt their prey. Because they are so large, their prey includes invertebrates plus small mice and other vertebrates, including smaller giant salamanders and their eggs. Little is known about their mating behavior because they get it on in underwater nooks and crannies among rocks. Males deposit spermatophores that females pick up with their cloacae, then females lay the fertilized eggs on the undersides of submerged rocks. Mating and egg laying likely occur in the spring, and incredibly the females then stay

California Giant Salamander. *Photograph by Marisa Ishimatsu.*

with the eggs until they hatch in the winter (possibly as protection from other giant salamanders intent on eating them!). The small aquatic larvae grow and feed on invertebrates and tadpoles for at least a year and often longer. In some cases, the larvae stay in the stream and become reproductively mature without otherwise metamorphosing.

Range and Variations: The California Giant Salamander is found in Santa Cruz and San Mateo Counties and extreme western Santa Clara County, and north of the San Francisco Bay in Marin, Sonoma, Napa, and southern Mendocino Counties, plus small bits of Solano and Lake Counties. The Coastal Giant Salamander's range picks up where the other species' range ends, ranging from southern Mendocino County northward through the coastal counties of California, Oregon, and Washington and into extreme southern British Columbia, Canada.

How to Find Giant Salamanders: Larvae are relatively easy to find in streams in appropriate habitat, as they tend to be abundant. Walk along the edge of streams in forested habitat looking for small larvae swimming at the edges and larger larvae in deeper areas. Adults are rarer, so you'll often need to put in some time and effort to find one. Flipping cover objects like rocks and logs in moist areas near streams can yield an adult giant salamander sheltering underneath. You might also get lucky like I did on that camping trip to the Santa Cruz Mountains and find an adult out walking around on the forest floor on a wet day in the late fall. I have also found them by driving around remote roads in their forest habitat on rainy nights.

Protection:
California Giant Salamander: species of special concern (California)
Coastal Giant Salamander: none

Clouded Salamander. *Photograph by Ryan Sikola.*

CLOUDED AND WANDERING SALAMANDERS

ANEIDES FERREUS AND *A. VAGRANS*

FAMILY PLETHODONTIDAE

Buckle into your harness and get your carabiners ready—this one is a wild ride. I've written in other parts of this book about the bias we have as scientists and amphibian enthusiasts based on our human senses, where we can see well but fail to detect most of the scents that are so important to amphibians. We also have difficulty observing what happens underwater and underground, both very important habitats for amphibians. But the closely related Clouded Salamander and Wandering Salamander occupy yet another niche that has been, until recently, inaccessible to us humans: the canopies of giant trees. Sure, one can find these salamanders hiding in the bark of logs on the ground. But just in the past twenty years, pioneering tree scientist Stephen Sillett of Cal Poly Humboldt became the first person to visit the canopies of giant redwood trees in Northern California using superhuman techniques involving archery and "skywalking" . . . but I digress, as this is a book on amphibians.

Sillett's goals in climbing these giants were to accurately measure their heights and understand how water gets transported over three hundred feet up their trunks, but when he arrived at the tops of these unexplored islands in the sky, he also discovered whole communities of plants and animals thriving there. In an NPR interview in 2010, Sillett said, "The most freaked out I have ever been was when we found the first salamander up there. And I am three hundred feet up in a tree and there is a salamander cruising around." I personally find it funny that a salamander is the thing that freaked out a person clinging to a tree by mere ropes three hundred feet high in the air, but that's neither here nor there. The salamander he found was a Wandering Salamander, which subsequent studies showed are abundant in the crowns of giant redwoods in California. The thick ferns and mosses on the branches harbor moisture year-round, so the salamanders never need to leave the trees. They do, however, sometimes move from tree to tree, but they don't have to descend them to do so. Rather, they jump and glide in their own form of skywalking. That's right, the big feet and long tails of these salamanders not only help them in grasping as they climb, but the salamanders also spread their feet and wiggle their tails as they glide from one big tree to the next. Even their toes play a role—a recent study showed that blood rushes into these salamanders'

Clouded Salamander. *Photograph by Francesca Heras.*

toe tips right before they jump off a tree, helping them release their grip and then stick their landing. So, if you think that every inch of the Earth's surface has already been explored, think again. What amphibians and other amazing critters occupy other unexplored places on our planet?

Appearance: These salamander species are medium in size with slim bodies, long legs, big feet, and square toe tips. Clouded Salamanders (*Aneides ferreus*) get their names from the cloud-like pattern of pale flecks and blotches on their brown or gray skin, which is similar in Wandering Salamanders (*Aneides vagrans*). The best way to distinguish them is based on location, as their area of overlap is only about fifteen kilometers and may consist of hybrids.

Natural History: These salamanders inhabit northern coastal forests, where they live inside and on trees and logs as well as under rock rubble. They can be found in high numbers at forest edges, including logged areas. They hunt for their arthropod prey on wet nights, and they also likely hunt underneath debris. Breeding occurs in spring. An adorable mating behavior has been described in Clouded Salamanders and is likely also exhibited by Wandering Salamanders. The female initiates a "circular tail-straddling behavior" where each member of the pair bends sideways while straddling the tail of the other, with one's head on top of its mate's hips. Both sexes are active participants in this courtship shuffle, rubbing their chins on one another, likely secreting pheromones that play a role in chemical communication. This is followed by the male performing an elaborate dance called the "foot shuffle," then depositing a sperm packet on the ground, which the female picks up with her cloaca. After fertilization, females lay eggs in crevices in wood or rocks and often tend their eggs until they hatch into tiny salamanderlings several months later.

Range and Variations: The Clouded and Wandering Salamanders were considered the same species until 1999, when genetic evidence supported separating them into distinct species. Clouded Salamanders occupy extreme northwestern California in Del Norte and Siskiyou Counties, and from there extend north throughout western Oregon. Wandering Salamanders are endemic to northwestern California, except for a long-term population on Vancouver Island in Canada that was likely introduced accidentally via shipments of bark used by artisans to tan leather.

How to Find Clouded and Wandering Salamanders: Like most plethodontid salamanders, the best method for finding Clouded

Wandering Salamander. *Photograph by Spencer Riffle.*

and Wandering Salamanders is to lift logs, small rocks, and other cover objects in areas where the ground is moist. Both can be found readily at forest edges where recent fires or logging activity have led to accumulated debris on the forest floor that provides excellent habitat. On wet nights, these salamanders can also be observed walking around on the surface.

Protection: none

Speckled Black Salamander. *Photograph by Spencer Riffle.*

BLACK SALAMANDERS

ANEIDES FLAVIPUNCTATUS, A. IECANUS,
A. KLAMATHENSIS, **AND** *A. NIGER*

FAMILY PLETHODONTIDAE

Black salamanders? Well, that's cool. Indeed, some members
of this group are a rich, pure black. Others have light flecks or
spots on their bodies. It's a mystery why some are solidly colored
while others look like they got between Jackson Pollock and his
canvas. Just like in California, some lineages of European newts
in the genus *Salamandra* are pure black while others have yellow
spots of various sizes. Intensive research has shown that the
pure black salamanders in this group have evolved multiple times
from more colorful ancestors. The driving force for this could be

thermoregulation, where darker individuals can absorb more rays from the sun. But whereas the European newts live at high elevations and may need to heat up to help warm their developing fetuses (they give birth to live young—how cool is that?), our black salamanders generally shun the sun and skulk around under rocks during the day, only coming out at night to feed. It's possible that pure black salamanders like the Santa Cruz Black Salamander may blend into the dark streamside rocks, helping them hide from predators and ambush prey. The speckled skin of the other three species may reflect light in a similar way as the wet leaves nearby, also upping their stealth. This whole idea is complicated by the fact that salamander predators and prey don't see the colors of the world the same way we do. Nocturnal species may see fewer colors but can detect very dim light, and lots of animals can see the ultraviolet light that is invisible to us. Adding to the fun is that even the completely black Santa Cruz Black Salamanders have babies with whitish-green speckling. Folks, if you came here for answers, I don't have them. I'm a scientist—I have loads of questions, a few hypotheses up my sleeves (see my skulk-and-stealth theory above), but one of you is going to need to go to graduate school and sherlock the heck out of "The Problem of the Black Salamander's Colors" as your own scientific question to solve.

Appearance: Black salamanders are medium in size with round toes that may help them in gripping rocks. The four species of black salamanders vary quite a bit in their coloration. The Santa Cruz Black Salamander (*Aneides niger*) is typically truly a solid black color, though juveniles can have a whitish-green sheen made by tiny spots all over their backs. The Speckled Black Salamander (*Aneides flavipunctatus*) can be solid black but more often is covered with white or yellow flecks or spots, similar to the Shasta

Clockwise from top left: A "frosted" variety of Speckled Black Salamander, Shasta Black Salamander, Santa Cruz Black Salamander, Klamath Black Salamander.
Photographs by Spencer Riffle.

Black Salamander (*Aneides iecanus*), which has white flecks all over its back. The Klamath Black Salamander (*Aneides klamathensis*) has more diffuse spots, described by scientists as "frosting," that give it a shiny, grayish-green hue. The eggs are laid in hidden spots you are unlikely to find, and the hatchlings look like tiny versions of adults.

Natural History: Black salamanders live mainly in wooded areas or grasslands. Some populations are associated with rocky streams. They hide under rocks or moist logs during the day and emerge at night to forage on small invertebrates. Little is known about their mating systems, but males appear to defend territories, likely based on access to females. Courtship occurs in the spring, when males deposit sperm packets on the ground that females pick up with their cloacae. In summer, females lay eggs by attaching them to the undersides of rocks or earth in hidden burrows, then stay with the eggs until they hatch into tiny salamanderlings in the fall.

Range and Variations: The four species of black salamanders in California were, until recently, considered to be subspecies of a single species. The Shasta Black Salamander is endemic to Shasta County. Also endemic to California is the Santa Cruz Black Salamander, which lives in Santa Cruz, San Mateo, and Santa Clara Counties and is listed as a California species of special concern due to human impacts on its fragile habitat in this area. Speckled Black Salamanders occur from Sonoma County north to Humboldt County, and the Klamath Black Salamander is found from Humboldt County north into southern Oregon.

How to Find Black Salamanders: The best way to find black salamanders is to look for them foraging near streams on wet nights or to flip cover objects. In some areas, the best cover objects are small rocks lining the edges of streams; this is especially true for most populations of Santa Cruz Black Salamanders. In other areas, black salamanders can be found hiding under rocks during the day much farther away from streams. For the three northern species of black salamanders, you can flip rocks alongside roads or clearings in redwood forests, as they seem to like being in areas with open canopies.

Protection:
Speckled Black Salamander: none
Shasta Black Salamander: none
Klamath Black Salamander: none
Santa Cruz Black Salamander: species of special concern (California)

Ryan Sikola

ARBOREAL SALAMANDER

ANEIDES LUGUBRIS

FAMILY PLETHODONTIDAE

Arboreal Salamanders are high on the most wanted list of students who take my Cal Poly herpetology class. They examine preserved specimens in lab and only sometimes get to see wild specimens in the flesh. Here on the central coast, Arboreal Salamanders are probably rather common but are not commonly seen, due to their tendency to hide in rocky crevices or under slabs of tree bark unless it is raining. Sure, they can sometimes be found underneath logs and can occasionally be seen walking around on the forest floor, but their prime time to be out galivanting is right smack in the middle of a rainstorm. Because it so rarely rains in my area of California, especially during the months of April through

June when I teach herpetology, it stands to reason that we don't see these salamanders very often.

When we do get lucky enough to see an Arboreal Salamander, my students are instantly smitten. I recall one student oohing and aahing over an Arboreal Salamander's adorable, grin-like visage, until the little beast bared its sharp teeth and emitted a loud squeak! This particular salamander was a large male with a huge, muscular head, and it could have packed a decent bite if the student hadn't hurriedly placed it back on its tree. Upon closer inspection, the salamander was covered in thin, U-shaped marks, which I explained to the students were scars from tussles with other males. As my student put it, "That salamander has stories to tell."

Wouldn't it be fun if salamanders could actually tell us their stories? Arboreal Salamanders can live for ten-plus years, inevitably racking up hundreds of nighttime adventures stalking bugs in the rain, guarding females from other males, and narrowly avoiding

Adult male Arboreal Salamanders have large, muscular heads. *Photograph by Ryan Sikola.*

Several juvenile Arboreal Salamanders found under the same cover object.
Photograph by Max Roberts.

the strikes of hungry snakes. These stories could fill up their own book—I like picturing them tucking away tiny, little waterproof field notebooks into crevices after a particularly wild night.

Appearance: Arboreal Salamanders are medium sized and pinkish-brown with light flecks or spots, with prominent grooves on their torsos, square toes, and large heads, especially in males. The salamanderlings are darker in color with speckling that looks like the Milky Way.

Natural History: Arboreal Salamanders live mainly in oak woodland and pine forest but can be found in other habitats as well, including rocky walls in suburban areas. They are nocturnal, emerging from moist hiding places to hunt on the surface, often on rainy nights. During courtship, males scrape their teeth along the females' backs to expose them to chemical secretions from glands on the males' chins, which may lead the females to pick up sperm packets the males place on the ground. Eggs are laid in the spring in moist areas like bark or rock crevices, then females stay with the eggs until they hatch into little salamanderlings. Arboreal Salamanders eat by ambushing invertebrates or small salamanders.

Range and Variations: Arboreal Salamanders are mainly a coastal species, ranging from Humboldt County down to Baja California. They also occur on the Farallon Islands, Año Nuevo Island, and a small part of the Sierra Nevada foothills east of the Bay Area.

How to Find Arboreal Salamanders: Arboreal Salamanders are lungless salamanders that breathe through their skin, so they are typically active when it is wet out. The best time to find them is right in the middle of a rainstorm. Shine your flashlight on rocky walls or oak trees at night to look for salamanders sitting on the surface or peeking out from crevices. You can also find them by flipping cover objects in the winter and spring. Sometimes you can find multiple salamanders under a single cover object, especially salamanderlings.

Protection: none

A tiny Black-bellied Slender Salamanderling. *Photograph by Spencer Riffle.*

Black-bellied Slender Salamander. *Photograph by Spencer Riffle.*

SLENDER SALAMANDERS

BATRACHOSEPS SPP.

This species account was very difficult to write. Why is that? Because I made the difficult decision to combine all twenty-plus species of California salamanders in the genus *Batrachoseps* into one account (hence the *spp.* in the title, meaning "multiple species"). The purpose of this book is to introduce amphibians to people who are new to the field of herpetology. All of these salamanders look similar to our eyes and have comparable natural histories, given that they are small lungless salamanders that remain underground or in moist hiding places during the dry season and emerge to feed and breed during the wet season. Though a professional herpetologist might recognize all the subtle differences among the slender salamander species, in this book separate accounts for each species would be a lesson in redundancy.

If all these salamanders look and act similarly, then why are they different species? Slender salamanders are tiny. While they may appear very similar to our eyes, there are big differences among species that are noticeable when you look closely. These species differ in characteristics like relative tail length, the size of the limbs, eye color, and the number of costal grooves (vertical grooves along each side of the body between the arms and legs). As genetic techniques improved over recent decades, the DNA of slender salamanders revealed a long and complex history of diversification in California. Several famous herpetologists, including Robert Stebbins, David Wake, Robert Hansen, and Elizabeth Jockusch, have focused heavily on slender salamanders. Wake, in particular, made the study of slender salamanders his life's work and was recently honored when the latest species of slender salamanders to be discovered was named after him (Arguello Slender Salamander, *Batrachoseps wakei*).

So, while slender salamanders appear very similar from a casual glance, they are diverse when examined in detail. Those of you interested in more information on slender salamanders in California should consult CaliforniaHerps.com or the field guide *California Amphibians and Reptiles*, by Robert Hansen and Jackson Shedd.

Getting back to our casual account, these little salamanders are collectively one of the most common and easiest amphibians to find in California. In many places, flipping any random log during the winter or spring is likely to reveal a slender salamander hiding under it. Many people dig them up while gardening and mistake them for little snakes, earthworms, or millipedes because their bodies are long and slender and their limbs are so tiny that

A Gabilan Mountains Slender Salamander coiled up alongside a millipede.
Photograph by Spencer Riffle.

they can be missed. They also have a fascinating tendency to coil up like millipedes when disturbed. This could potentially be a form of mimicry for deterring predators because millipedes secrete a nasty cocktail of toxins when harassed, including chemicals like cyanide and hydrochloric acid.

Their secrecy makes it difficult to estimate how many slender salamanders occupy a given area. That is too bad because I bet the biomass of these salamanders in California is huge. They therefore play a major role in California food webs by eating millions of tiny invertebrates and turning them into salamander biomass, which can then be eaten by snakes, birds, and large insects and arachnids.

Appearance: Slender salamanders have elongated bodies with long tails, small arms and legs, and bulbous eyes. The coloration of slender salamanders varies widely, both among and within species. Common color patterns include solid dark brown, mottled gray or tan, and a reddish or light brown stripe down the back and tail. All slender salamanders have four toes on their hind feet, whereas other salamanders in California have five toes. Consult a field guide for the specific physical differences among the many species of slender salamanders in California.

Natural History: Species with large ranges, like the California Slender Salamander, can be found in a variety of habitats, from forests to grasslands to urban gardens. Because of their small size, the skin of slender salamanders dries out rapidly, which could be deadly to an animal that needs its skin to be moist to breathe across it. They therefore are found in moist microhabitats, like under logs, rocks, and leaf litter. Species with tiny geographic ranges may be found in more specific habitat types, like rocky desert canyons, as in the case of the Inyo Mountains Salamander. During the dry season, they move deeper underground and hide out until rains arrive. For a genus of salamanders whose evolutionary relationships have

California Slender Salamander. *Photograph by Marisa Ishimatsu.*

Inyo Mountains Salamander. *Photograph by Brandon Kong.*

been studied so extensively, there is remarkably little known about their reproductive biology. They likely mate like other plethodontid salamanders, with the male leaving a sperm packet on the ground that the female picks up with her cloaca. She then lays eggs in a moist area, like under a log; in some species multiple females lay eggs in a communal nest. Unlike other plethodontid salamanders, the females do not stay and protect the eggs. Mating and egg laying typically take place in fall or winter following the first rains, and the eggs hatch into tiny salamanderlings later that winter. Slender salamanders eat tiny terrestrial invertebrates.

Range and Variations: Here, I list all of the species of slender salamanders that occur in California with a general description of where they occur and their protected status (if applicable):

Greenhorn Mountains Slender Salamander (*Batrachoseps altasierrae*): western side of Sierra Nevada in Tulare and Kern Counties. Protection: none.

Desert Slender Salamander (*Batrachoseps aridus*): small area of Santa Rosa Mountains in Riverside County. Protection: endangered (California and federal).

California Slender Salamander (*Batrachoseps attenuatus*): coastal counties from the Bay Area north to southwestern Oregon, foothills of northern Sierra Nevada, and isolated populations in northern Central Valley. Protection: none.

Fairview Slender Salamander (*Batrachoseps bramei*): small area of Kern River Canyon in Tulare and Kern Counties. Protection: none.

Inyo Mountains Salamander (*Batrachoseps campi*): Inyo Mountains, Inyo County. Protection: species of special concern (California).

Hell Hollow Slender Salamander (*Batrachoseps diabolicus*): western side of Sierra Nevada from Mariposa County north to El Dorado County. Protection: none.

San Gabriel Mountains Slender Salamander (*Batrachoseps gabrieli*): San Gabriel and San Bernardino Mountains in Los Angeles and San Bernardino Counties. Protection: none.

Gabilan Mountains Slender Salamander (*Batrachoseps gavilanensis*): Occurs in a thick stripe from Santa Cruz County southeast to northern San Luis Obispo and Kern Counties. Protection: none.

Gregarious Slender Salamander (*Batrachoseps gregarius*): western side of the Sierra Nevada from Mariposa County south to Kern County. Protection: none.

San Simeon Slender Salamander (*Batrachoseps incognitus*): Santa Lucia Mountains in southern Monterey County and northern San Luis Obispo County. Protection: none.

Sequoia Slender Salamander (*Batrachoseps kawia*): western side of Sierra Nevada Mountains in Tulare County. Protection: none.

Santa Lucia Mountains Slender Salamander (*Batrachoseps luciae*): Santa Lucia Mountains in Monterey County. Protection: none.

Southern California Slender Salamander (*Batrachoseps major*): coastal areas from Los Angeles County south to Baja California, Mexico, plus Santa Catalina Island. Protection: none.

Lesser Slender Salamander (*Batrachoseps minor*): Santa Lucia Mountains in San Luis Obispo County. Protection: species of special concern (California).

Black-bellied Slender Salamander (*Batrachoseps nigriventris*): coastal counties from Monterey County south to Orange County, plus Tehachapi Mountains in Kern County and Santa Cruz Island. Protection: none.

Channel Islands Slender Salamander (*Batrachoseps pacificus*): Anacapa, San Miguel, Santa Cruz, and Santa Rosa Islands. Protection: none.

Kings River Slender Salamander (*Batrachoseps regius*): western side of Sierra Nevada in Fresno County and extreme northern Tulare County. Protection: none.

Relictual Slender Salamander (*Batrachoseps relictus*): Breckenridge Mountain in Kern County. Protection: species of special concern (California), with federal endangered status currently pending.

Kern Plateau Salamander (*Batrachoseps robustus*): Kern Plateau in Sierra Nevada, Tulare County, and eastern side of Sierra Nevada in Inyo County, plus a tiny area of northern Kern County. Protection: none.

Kern Canyon Slender Salamander (*Batrachoseps simatus*): Kern River Canyon, Kern County. Protection: threatened (California), with federal threatened status currently pending.

One of the only photos in existence of the newly described Arguello Slender Salamander, whose tiny range lies entirely within Vandenberg Space Force Base. *Photograph by Brandon Kong.*

Tehachapi Slender Salamander (*Batrachoseps stebbinsi*): Sierra Nevada and Tehachapi Mountains, Kern County. Protection: threatened (California).

Arguello Slender Salamander (*Batrachoseps wakei*): tiny area on Vandenberg Space Force Base at Point Arguello, Santa Barbara County. Protection: none.

Undescribed species: Several potential new species of slender salamanders have been recently discovered but are still being studied and do not yet have a scientific name. These include the Borrego Slender Salamander from Borrego Springs in San Diego County, the Peninsular Slender Salamander from southern Imperial County and likely extending into northern Mexico, and potentially additional species by the time this book is published.

How to Find Slender Salamanders: Finding slender salamanders involves one main method: flipping. Head to appropriate habitat during the right time of year (the soil should be moist, like during the rainy season in winter and spring in most areas of the state), and flip cover objects like logs and flat rocks. Sometimes you can find multiple slender salamanders under the same log. I think the record number of Black-bellied Slender Salamanders that my students have seen under one log is fifteen! Be sure to gently pick up the salamander or scoot it out of the way before replacing the cover object into its exact outline, then let the salamander go right next to the edge so it can crawl to safety. Slender salamanders are active on the surface sometimes, typically at night when it is wet outside, but they are so small that it can be difficult to see them.

Protection: see Range and Variations

Monterey Ensatina. *Photograph by Spencer Riffle.*

ENSATINA

ENSATINA ESCHSCHOLTZII

FAMILY PLETHODONTIDAE

The Ensatina is an iconic Californian salamander. Even though it ranges from southern Canada to northern Mexico, its range inside California has made it famous well beyond the realm of herpetology. Ensatina is a star in many evolution textbooks as an example of a "ring species," where various subspecies occur in a ring distribution, each interbreeding with its neighbor, but two of the subspecies at each end of the ring cannot interbreed. In the case of Ensatina, they range in a ring around the coast, Northern California, and the Sierra Nevada, and the two subspecies at the southernmost part of the range are the ones that don't interbreed (at least not readily). The reason ring species are important in the study of evolution is that

This Oregon Ensatina is exhibiting a defensive posture and secreting a sticky goo to deter a predator. *Photograph by Spencer Riffle.*

they challenge the most basic definition of a species: if members of the same species cannot breed with one another, are they truly members of the same species? There is no good answer to that, with scientists falling into different camps about the definition of a species and how ring species fit into that. The very idea that Ensatina is a ring species has even been disputed and dissected relentlessly over the years, all of which has led to advancements in evolutionary biology but little clarity on the complex relationships among the brightly colored variants in this species.

One thing that is not in dispute is Ensatina's formidable (and adorable) means of defending itself against predators (and curious humans). The name *Ensatina* means "sword-shaped" in Latin, and the reason for this becomes clear when you witness a harassed Ensatina pointing its tail at you as if saying, "En garde!" But the threat of this sword lies not in being pointy but rather in the sticky substance that it emits when handled by a predator.

The tail can also easily drop from the body, where it proceeds to wiggle around. Imagine a predator grabs an Ensatina by the tail. It will soon find itself with a mouthful of squirmy flesh emitting a sticky goop that glues its jaws closed while the Ensatina sidles away into the brush. Fascinatingly, scientists have not yet studied whether the secretions are merely sticky or whether they also taste bad or even are toxic to predators—this would make a great student research project!

These beautiful salamanders can be easy to find in the right areas at wet times of year, making hunts for them satisfying and appropriate for kiddos. But beware when Ensatina hunting—not of their toxic tails but rather of the oils of the poison oak plant with which they so frequently share habitat. Put on thick pants before heading out to the understory on your hunt, watch where you put your hands, and, as always, put the Ensatinas' roofs back exactly as you found them.

Sierra Nevada Ensatina. *Photograph by Marisa Ishimatsu.*

Appearance: Ensatinas are medium-sized salamanders with large, bulging eyes and a characteristic constriction at the base of their tails. Their coloration varies widely based on geographic location (see figure below).

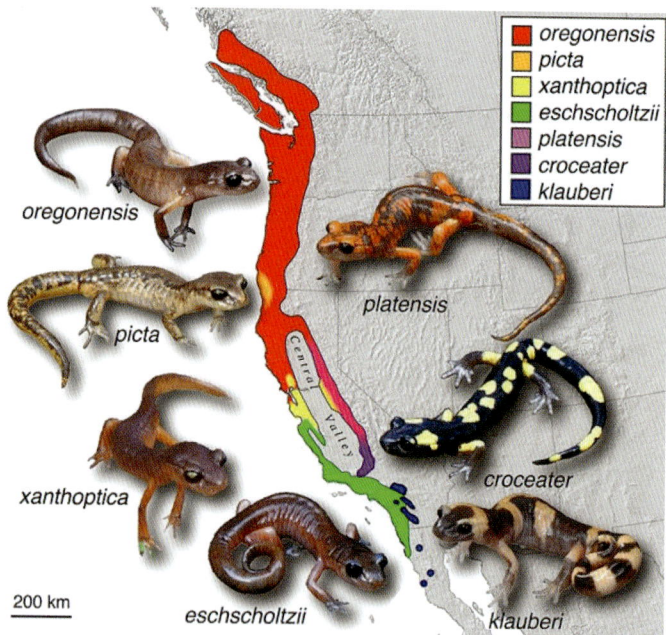

This map shows all of the subspecies of Ensatina. *Photograph from Thomas J. Devitt, et al., "Asymmetric Reproductive Isolation Between Terminal Forms of the Salamander Ring Species* Ensatina eschscholtzii *Revealed by Fine-Scale Genetic Analysis of a Hybrid Zone" (BioMed Central Ltd, CC BY 2.0).*

Natural History: Ensatinas are terrestrial salamanders, spending most of their time hiding in moist areas like little tunnels within the soil and underneath cover objects like logs. They are found

Some scientists consider the Large-blotched Ensatina to be a separate species from the others. *Photograph by Marisa Ishimatsu.*

mainly in forested areas with plenty of fallen wood and other cover. In wet conditions, these salamanders will emerge into the open to hunt, usually at night. Like all salamanders in the family Plethodontidae, Ensatinas lack lungs, so they keep their skin moist to facilitate gas exchange. To mate, Ensatinas engage in a cute courtship dance where the male sits in front of the female and winds his tail around hers, then deposits a package of sperm onto the ground and helps scoot the female over it so she can pick it up with her cloaca. Mating happens in fall or spring, and the female lays her eggs under a cover object or in a soil tunnel. She stays with the eggs and guards them until they hatch in the late summer or fall. Ensatinas eat many kinds of arthropods and other invertebrates by grabbing them with their sticky tongues and pulling them into their mouths.

Range and Variations: Ensatinas range from southwestern Canada to northwestern Baja California, Mexico. In California, they occur in a ring distribution, as shown in the figure on page 80. Note that these geographic boundaries are approximate, and the subspecies often interbreed at the borders of their ranges. In coastal Southern California and northern Mexico, the Monterey Ensatina (*Ensatina eschscholtzii eschscholtzii*) is found, while in some inland mountain ranges you can instead find the Large-blotched Ensatina (*Ensatina eschscholtzii klauberi*). These are the two subspecies that do not readily interbreed, and so some people think the Large-blotched Ensatina is actually a separate species. Further north near the Bay Area is the Yellow-eyed Ensatina (*Ensatina eschscholtzii xanthoptica*), which can also be found in the western foothills of the Sierra Nevada. North of the bay you will find the Oregon Ensatina (*Ensatina eschscholtzii oregonensis*), which goes all the way up to Canada, except for a small slice of the coast at the California–Oregon border where the Painted Ensatina (*Ensatina eschscholtzii picta*) occurs. The Sierra Nevada Ensatina (*Ensatina eschscholtzii platensis*) occurs from the Mount Lassen area south to Kern County, where they give rise to the Yellow-blotched Ensatina (*Ensatina eschscholtzii croceater*).

How to Find Ensatinas: Ensatinas are superbly common in some areas and rare in others. Take two California Polytechnic State University campuses as examples. On and near the campus of Cal Poly Humboldt, Ensatinas are by far the most common salamander, often with multiple individuals under a given log. In contrast, Ensatinas are rare in natural areas on and around the campus of my university, Cal Poly San Luis Obispo. The best way to find Ensatinas is to look for them under logs and other cover objects during times of the year when the ground is wet, including spring. In areas with mild weather, winter is also an excellent time to search for them. On warm rainy nights, you might find them out and about by hiking around with a flashlight.

Protection: none

Mount Lyell Salamander. *Photograph by Zeev Nitzan Ginsburg.*

WEB-TOED SALAMANDERS

HYDROMANTES BRUNUS AND *H. PLATYCEPHALUS*

FAMILY PLETHODONTIDAE

Despite living in California for much of my life, I didn't visit Yosemite National Park until I was forty-five years old. This is partly due to my fear of crowds—I much prefer to take my wanderings off the beaten path by myself or with a small group of students and colleagues, happily peering under rocks in the wilderness. So, you will understand why I am so fascinated with the story of the discovery of the Mount Lyell Salamander in the park in 1915 by Charles Camp, an undergraduate researcher in UC Berkeley's Museum of Vertebrate Zoology, where I held the same position a mere eighty years later. As part of a survey of the park's mammals, Camp set a line of traps just downslope from the tree line at about 11,000 feet.

Mount Lyell Salamander. *Photograph by Marisa Ishimatsu.*

Imagine his surprise when he opened a trap and found not a familiar, furry rodent but rather a male and female salamander pair, the likes of which no biologist had seen before! In his 1916 publication describing his discovery, Camp rather anticlimactically wrote, "The two specimens were found to have been captured simultaneously in a spring-clip mouse-trap set in front of a small hole running into the moist soil beneath some rocks." Dry scientific prose aside, I like to think that when he realized he was looking at a brand-new species, he jumped for joy or maybe even danced a little old-timey jig right there on the slope of Mount Lyell, after which he ended up naming the species.

Camp had very little to compare these salamanders to, remarking only that they somewhat resembled another salamander species captured in an alpine area of the Mexican highlands. Fast forward thirty years to the middle of the twentieth century, and this salamander had been studied quite a bit more, but still very little was known about it. In *Reptiles and Amphibians of Yosemite National Park* (1946), Myrl Walker wrote, "Of all the forms of vertebrate life found in Yosemite National Park, probably no other is so well known to the scientific world, yet so little known to the average Park visitor, as this very strange species of salamander."

This same publication now gives a fascinating glimpse into how the park has changed in the past century. Unlike some other salamanders in the park, the Mount Lyell Salamander does not need to be near ponds or streams because it mates on land, and the water from melting snow and glaciers helps keep its skin wet. This means it can live in seemingly unlikely places, like on top of the huge, exposed mass of granite known as Half Dome. Walker wrote:

> A regular "colony" make their home under the flat rocks in the little valley on top of Half Dome, where the snow field lasts well into July, and where they are kept moist by the cold snow water seeping along in the soil beneath the flat rocks. More Mount Lyell salamanders have been collected from the top of Half Dome than from any other simile locality in Yosemite National Park.

If you are one of the lucky people to have hiked up Half Dome, you might know that camping is now forbidden at its upper reaches. The Mount Lyell Salamander is at least partly responsible for this. According to a sign posted at this iconic Yosemite landmark, campers were destroying salamander habitat to make rock shelters . . . and apparently to cover their "number twos." Rangers found that the flat rocks once favored by the salamanders harbored not the amphibians but rather piles of poop and toilet paper—over an area of five acres! Folks, that is *not* what the phrase "leave no stone unturned" means.

Mount Lyell Salamander. *Photograph by Spencer Riffle.*

In this account, I include both the Mount Lyell Salamander and the Limestone Salamander because they are both in the same genus and endemic to California, range in and near the Sierra Nevada, and are relatively unstudied and mysterious when it comes to their natural history.

Appearance: Both species are small salamanders that share a flattened body shape that allows them to sidle into rock crevices. Their webbed toes and short tails are adaptations for climbing on rocks. The Limestone Salamander (*Hydromantes brunus*) is brown on top and pale underneath, and the Mount Lyell Salamander (*Hydromantes platycephalus*) has a blotchy gray-and-brown pattern on top that helps it blend in with the granite rocks on which it lives, with a dark belly sometimes flecked with a silvery sheen.

Limestone Salamander. *Photograph by Ryan Sikola.*

Natural History: Limestone Salamanders and Mount Lyell Salamanders are rock-dwelling salamanders, almost always found in association with limestone and granite, respectively. Limestone Salamanders live among mossy rocks at low elevations in a specific river drainage (see Range and Variations), whereas Mount Lyell Salamanders live among wet rocks fed by snowmelt or waterfalls at higher elevations in the Sierra Nevada, plus creeks in parts of the Owens Valley. Very little is known about reproduction in either species. Like other salamanders in the family Plethodontidae, they likely mate when a female picks up a sperm packet deposited by a courting male. They probably lay eggs in hidden rock crevices in the late summer, and hatchling Mount Lyell Salamanders have been observed the following summer. They eat small insects and other invertebrates via a fascinating mechanism where they ballistically eject their "throat skeleton," essentially shooting their sticky tongues out to catch prey from as far as an entire body length away.

Range and Variations: The Limestone Salamander is endemic to California, living only in limestone rocks lining parts of the Merced River and its tributaries in Mariposa County. The Mount Lyell Salamander obviously occupies Yosemite National Park where it was first discovered but is also found in other mid- to high-elevation areas in the Sierra Nevada from Tulare County north to Sierra County, and on the eastern slope of the Sierra Nevada in the Owens Valley as far south as the Big Pine area. A possible third species related to the Mount Lyell Salamander was recently discovered in the Millerton Lakes Cave System in Fresno County, but this has not yet been officially described.

How to Find Web-Toed Salamanders: I have never looked for Limestone Salamanders in the wild, and I suggest you avoid them, too, to protect the sensitive habitat within their tiny geographic range. I've been told that they can most easily be found by peering into limestone crevices or flipping rocks along the Merced River during wet weather from fall to spring, as long as it is not too cold. Mount Lyell Salamanders can be found by flipping flat, granite rocks during the spring or summer when water is seeping under the rocks. They can also be found climbing on rocks within the spray of waterfalls on spring and summer nights.

Protection:
Limestone Salamander: threatened (California)
Mount Lyell Salamander: none

Samwel Shasta Salamander. *Photograph by Spencer Riffle.*

SHASTA SALAMANDERS

HYDROMANTES SAMWELI, H. SHASTAE, AND *H. WINTU*

FAMILY PLETHODONTIDAE

Each year in December, I drive back and forth across Shasta Lake to visit my family in Washington. Some years when it is wet and unseasonably warm, it is all I can do not to delay my arrival at the family holiday party by pulling over and hiking around looking for Shasta salamanders. These fascinating little creatures are found only in small populations surrounding this lake, which is actually a manmade reservoir that was created when the Shasta Dam was erected in the 1940s to supply water to the Central Valley. Dams always lead to trade-offs, with the benefit of water for farming often winning out over the health of wildlife populations and Indigenous people in the area. The flooding of the reservoir not only submerged a huge amount of habitat for the Shasta salamander species complex, but it also destroyed areas sacred to the Winnemem Wintu people and blocked the McCloud River migration route of the Chinook Salmon, a major food source for the tribe.

In 2018, what was previously considered a single salamander species was split into three species that look practically the same to our eyes but show important genetic divergence. The scientists named one of the species the Wintu Shasta Salamander (*Hydromantes wintu*) to honor the people whose Native lands and ways were devastated by the dam. The other new species was named the Samwel Shasta Salamander (*Hydromantes samweli*), after the nearby Samwel Cave, whose name was derived from "Sa-Wal," the Wintu word for the sacred Grizzly Bear. Populations on the south

Shasta Salamander. *Photograph by Spencer Riffle.*

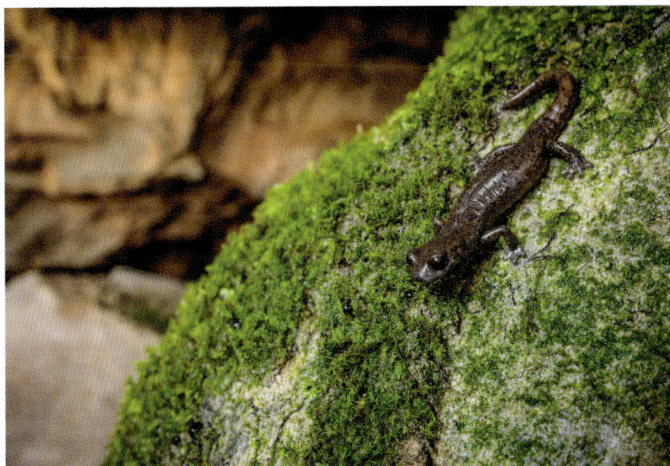

Samwel Shasta Salamander. *Photograph by Marisa Ishimatsu.*

and east sides of the lake remained known as the Shasta Salamander (*Hydromantes shastae*). The erection of the dam decades
ago was not the final chapter in this saga, however. As drought
severity and the population of hungry and thirsty Californians
increase, more and more water is needed in the Central Valley.
Many farmers want the dam's height to be increased to provide
water for crops, which would flood additional sacred Indigenous
sites and greatly impact the salamanders and other wildlife. Since
all three species of Shasta salamanders are protected under the
California Endangered Species Act, it is possible that the proposed
alteration of the dam might be blocked. At the time I write this
book, the bill containing the dam alteration (H.R. 215) is in political
and legal limbo.

Water access is a major problem that will likely only increase
in severity in California, so we herpetologists are watching with
bated breath to see whether today's government will repeat the

Wintu Shasta Salamander. *Photograph by Marisa Ishimatsu.*

trade-offs of the past or instead choose to prioritize biodiversity and the rights of our state's Indigenous people. Encouragingly, in 2024 a series of dams in the Klamath River further north were removed, restoring the salmon run and an important food and cultural resource for the Yurok people for the first time in over a hundred years.

Appearance: All three species of Shasta salamanders are similar in appearance. They are small with short tails and large webbed feet to help them climb around on rock faces. They are brownish in color with mottled gold, silver, or greenish specks.

Natural History: These salamanders are found exclusively in limestone and volcanic rock outcrops and nearby forests in small areas surrounding Shasta Lake. They can be seen climbing around on rocks on wet nights and may be found under rocks during the day. Mating has not been observed in these species, but they are probably like other salamanders in their family in that the males deposit sperm packets on the ground that the females pick up during courtship dances. Females lay eggs in the late summer or fall, and these hatch into tiny salamanderlings later in the fall. Shasta salamanders feed on small invertebrates. Like other salamanders in the family and especially in the genus, they are able to capture prey from several inches away by ejecting their tongues ballistically.

Range and Variations: As described on the previous page, the three species of Shasta salamanders have tiny ranges near the shores of Shasta Lake. To find the exact ranges of the three species, consult a range map in a current field guide. A fourth species of salamander in the genus *Hydromantes* was recently discovered north of Lake Shasta in the vicinity of Castle Crags, but at the time of writing this book it has not yet been officially described.

How to Find Shasta Salamanders: You can find Shasta salamanders by hiking on rocky outcrops and cliffs on wet nights in the fall, winter, and spring and looking for the salamanders climbing around on the rock walls. Shine your flashlight into rocky crevices to look for hiding salamanders. During the day, you can try flipping small rocks. Please take extreme care not to injure them or destroy the habitat when searching for these salamanders. Because these salamanders are all listed as threatened by the state of California, it is not permissible to capture or even touch them.

Protection: all three are threatened (California)

Dunn's Salamander. *Photograph by Ryan Sikola.*

WOODLAND SALAMANDERS

PLETHODON ASUPAK, P. DUNNI, P. ELONGATUS, AND *P. STORMI*

FAMILY PLETHODONTIDAE

There is a living puzzle in extreme Northern California. It hides out among the moss-covered rocks of old-growth forests, waiting for a dedicated graduate student to come and piece it together. Though previous scientists have tried, the puzzle of the taxonomic status of California woodland salamanders in the genus *Plethodon* is far from being solved. Some consider three of these species (Scott Bar, Del Norte, and Siskiyou Mountains Salamanders) to be subspecies of a single species, related to their "sister" species the Dunn's Salamander. Most consider them separate species, as I show here. The Dunn's Salamander barely dips from Oregon into

California, and the other three species are endemic to California and occupy tiny geographic ranges, making me wonder whether more sampling in the remote forests of the California–Oregon borderlands will yield additional puzzle pieces. Furthermore, little is known about the ecology and behavior of these salamanders. For these reasons, I have merged them into a single account here, where I compare and contrast their appearances and what little is known about their ecologies, and provide additional resources for those interested in snapping a puzzle piece into place.

One more thing unites all of these salamanders: concern over their well-being. First, living in remote areas where few people visit does not protect animals from the dangers of climate change. What might appear remote on a map may actually be a bustling center of logging activity, which is certainly true in many parts of the mountains of Northern California and Oregon. Another factor is the tiny areas occupied by these species, especially the Scott Bar and Siskiyou Mountains Salamanders, both of which have geographic ranges much smaller than most people's daily commutes. The third factor is one that involves you, should you decide to go

Scott Bar Salamander. *Photograph by Spencer Riffle.*

Dunn's Salamander. *Photograph by Ryan Sikola.*

search for these salamanders. They all live among moist, mossy rocks that lie beneath the tree canopy. Flipping these rocks or even walking around on them can destroy habitat and kill salamanders. If you visit the special places that these salamanders call home, be sure to tiptoe around the edges of the rocks and leave *all* stones unturned.

Appearance: All three of these salamanders are medium sized with narrow bodies and grooves on their sides. Here, I provide general descriptions of their appearance. The Dunn's Salamander (*Plethodon dunni*) has typical salamander toes, while the other three species have partially webbed toes. Dunn's Salamanders usually are dark brown with light blotches on their sides and a rough-edged, lighter stripe down their backs that does not go all the way down the tail, giving them a distinct dark tail tip. The Scott Bar Salamander (*Plethodon asupak*) is brown with tiny, light flecks that are concentrated along its sides and belly. The juveniles often have two orange stripes on their backs that fade with age. The Del Norte Salamander (*Plethodon elongatus*) is also brown but usually

Del Norte Salamander. *Photograph by Spencer Riffle.*

has a thick, reddish stripe down its back, with some flecks along its sides and belly. Juveniles have bolder red stripes down their backs. The Siskiyou Mountains Salamander (*Plethodon stormi*) is—you guessed it—brown, but usually has more intense flecking. Juveniles vary in color but often have lighter stripes on their backs and dark bellies.

Natural History: These salamanders live in the rocky rubble that forms the forest floor in their ranges. They shelter underneath the rocks during extreme cold and heat, and they hunt for prey among the verdant mosses that provide water and nutrients to a wealth of invertebrates. Research has shown that the Del Norte Salamander can also sometimes be found hiding under logs or rocks in dry riverbeds, and the other species may do so as well. The reproductive biology of these salamanders is undescribed, so we can only assume that they follow the typical plethodontid pattern whereby females pick up sperm packets that males deposit on the ground for them during courtship rituals, then the females lay eggs in the spring among the nooks and crannies of their rocky habitats and

Siskiyou Mountains Salamander. *Photograph by Spencer Riffle.*

tend them until they hatch in the fall. Dunn's Salamanders apparently can reproduce year-round, since females with eggs have been observed in all seasons.

Range and Variations: The Dunn's Salamander ranges from southern Washington through western Oregon and barely makes it into Del Norte County in extreme northwestern California. The Siskiyou Mountains Salamander is found in a small area of the northern part of its namesake Siskiyou Mountains, and from there extends a short distance into southern Oregon. The Scott Bar Salamander has the smallest geographic range of any northwestern salamander. It lives in the mountains slightly to the east of the Siskiyou Mountains Salamander, near the confluence of the Scott and Klamath Rivers. The Del Norte Salamander occupies areas of Del Norte County, northern Humboldt County, western Siskiyou County, and a small slice of northwestern Trinity County, and also extends into southwestern Oregon. More specific information on the geographic ranges of these salamanders, especially for those planning trips and needing vehicle access points to find them, can be found in a field guide or by consulting iNaturalist.

How to Find Woodland Salamanders: The most effective way to find these salamanders is unfortunately an irresponsible method because it risks destroying the homes of these sensitive species. Flipping the moss-covered rocks under forest canopies and at the edges of forests can yield salamanders hiding underneath. If you must flip rocks, please choose easily flippable rocks along the edges and take care to replace them exactly as found. The salamanders can also be found on the surface on rainy nights in the spring.

Protection:
Scott Bar Salamander: threatened (California)
Dunn's Salamander: species of special concern (California)
Del Norte Salamander: species of special concern (California)
Siskiyou Mountains Salamander: threatened (California)

SOUTHERN TORRENT SALAMANDER

RHYACOTRITON VARIEGATUS

FAMILY RHYACOTRITONIDAE

While working on this book, as I have been enjoying chasing winter storms all over Northern California in search of one cool salamander after another, I have realized how special salamanders are. The family Rhyacotritonidae is *really* special. It has only one genus (*Rhyacotriton*), with four species, all of which occur in the coastal Pacific Northwest and just one in California. For many years, scientists argued over where this family fit into the salamander family tree, until just recently when its place as a close relative of the family Plethodontidae was confirmed using modern genetic techniques. Well, "close relative" may be a bit of a stretch. Scientists estimate that the common ancestor of Rhyacotritonidae and Plethodontidae split off nearly 150 million years ago. Today, the descendants in each family still share many similarities but have also become different. Both groups of salamanders have drastically reduced lungs such that most of their gas exchange occurs across their skin or the lining of their mouths. However, plethodontid salamanders are fully terrestrial and would drown if placed in the water, and rhyacotritonids like the Southern Torrent Salamander are mostly aquatic. Courtship and mating rituals take place on land in both groups, but plethodontid eggs are laid in moist places on land and hatch into terrestrial salamanderlings, whereas the Southern Torrent Salamander lays its eggs underwater where they hatch into tiny gilled larvae.

The Southern Torrent Salamander relies on clean, cold-water streams in old-growth forests for all aspects of its life, from feeding to breeding. It therefore follows that the destruction of these forest habitats by logging has dramatically impacted these special little salamanders. The Southern Torrent Salamander is protected from capture in California, but people's demand for lumber to frame their houses, shingle their roofs, build their railroads, and train their grapevines began ruining a huge amount of its habitat in the mid-nineteenth century, a trend that hasn't eased much today. Logging occurring upstream can increase silt runoff, which then clogs up the little cracks and crevices used by larval and adult salamanders. Without such hiding spots, they run the risk of being hunted down and made into dinner by Coastal Giant Salamanders (cute to us, terrifying to a wee torrent salamander). Until its habitat is fully protected and managed, scientists worry that populations will continue to go extinct and fragment this species.

Marisa Ishimatsu

Appearance: Southern Torrent Salamanders are medium in size, with short tails and bulging eyes. They are dark green or brownish on top and their bellies are yellow, and both sides have dark and light speckling. Their toe tips are noticeably squared off. Males of all members of this genus have distinct square-shaped lobes of skin framing their cloacae. You are unlikely to ever see their eggs, which are cream-colored and laid singly, attached to the bottoms of underwater rocks. The small aquatic larvae have tiny external gills and finned tails.

Natural History: Southern Torrent Salamanders, like all members of this family of salamanders, are found in shallow, rocky streams running through old-growth coastal forests. They are mostly aquatic, found either in the cold water or hiding under small rocks on wet substrate at the stream edges. Mating behavior of these salamanders has never been observed in the wild, but scientists have put cameras on Southern Torrent Salamanders in the laboratory and sneakily watched them mating. One side of each container was filled with water so that salamanders could choose to be in the water or on "land." Courtship and mating always occurred at night, on the land section of the containers. Males perform an elaborate courtship dance by lifting their tails and performing a "tail-waggling display." Both salamanders rub one another with their chins, possibly transferring and sensing pheromones. The male deposits a sperm packet on land and helps lead the female, who is straddling his tail, to pick it up with her cloaca. In this study, the voyeur scientists had the most luck getting the salamanders to mate when they put more than one male in each container, where they apparently revved each other up. They regularly chased and bit one another, and sometimes a male even butted into an encounter between a couple and pretended to be a female salamander, distracting the other male long enough to sneakily deposit his own sperm packet! Following internal fertilization, the female heads into the water and

Ryan Sikola

lays eggs in hidden spots, attaching them singly to the bottom of rocks and inside crevices. Egg laying usually occurs in the summer, and the eggs do not hatch until the following spring. The tiny larvae are also slow to develop, feeding on small aquatic invertebrates and growing for several years before metamorphosing into adult form. The adults feed on small invertebrates including springtails and amphipods.

Range and Variations: Southern Torrent Salamanders occur in coastal areas of northwestern California from Mendocino County northward into coastal Oregon.

How to Find Southern Torrent Salamanders: On the one hand, Southern Torrent Salamanders have very specific habitat requirements, making them relatively easy to find if you know how to look for the right shallow, rocky streams in old-growth forests within their range. On the other hand, such streams are increasingly hard to come by and are very sensitive areas. Be sure to visit such areas with care, cleaning your boots before you arrive and avoiding treading on the little rocks that salamanders may hide beneath. The best way to find Southern Torrent Salamanders is to flip small and medium-sized rocks lying along the edges of these streams. If you find a salamander underneath one, you might be amazed at how quick they are—when disturbed, they will suddenly thrash around violently and escape into the water before you know it. If you flip a rock, you should carefully replace the rock into its exact outline (gently nudge the salamander out of the way if necessary, then allow it to crawl back under the rock).

Protection: species of special concern (California)

ROUGH-SKINNED NEWT

TARICHA GRANULOSA

Curl up and get ready for a good story that I like to call "Revenge of the Newt." Although no one knows for sure how much of what I'm about to relay to you is fact versus fiction, I can attest that the biology behind it is 100 percent true. The story goes that a group of friends went on a hunting and camping trip in the Pacific Northwest. When they didn't return home on time, their families got worried and a search party was formed. The campsite was found with nothing initially appearing awry . . . except that the men were all lying deceased inside their tents.

Investigators allegedly found a dead Rough-skinned Newt inside the coffee pot, leading them to hypothesize that the salamander had crawled into the pot undetected and was brewed along with the grounds at coffee hour, thereby poisoning the unwitting hunters. While the likelihood of such an occurrence is miniscule (don't let it scare you away from camping in newt country!), it highlights the dramatic toxicity of the Rough-skinned Newt. How could such a small animal be so poisonous?

All four species of newts in the genus *Taricha* in California, but especially Rough-skinned Newts, produce a deadly chemical called tetrodotoxin that can poison anything daring to ingest them by blocking impulses in their nerves. In some populations of Rough-skinned Newts, a single salamander has enough toxin to kill twenty people! Why on Earth would it be so poisonous? Clues began accumulating in recent decades as father-and-son scientist team Edmund "Butch" Brodie discovered that in some areas, especially those near the San Francisco Bay Area, garter snakes are able to eat highly toxic newts. These garter snakes have genetic mutations that make them resistant to tetrodotoxin, so they can take advantage of newts as an abundant food source that few other predators can tolerate. It's likely that an evolutionary arms race between the newts and snakes, where a more resistant snake could eat a more toxic newt, has led to natural selection for extremely high toxin levels in some populations of newts.

But that's not the full story. It turns out that a garter snake that eats a toxic newt can sequester the toxin in its liver for weeks, making itself toxic to any predator that would try to eat it! This incredible story of predator and prey involves many additional layers, with scientists still unraveling even the most basic components of the tale. For example, researchers still don't know for

Spencer Riffle

sure whether the newts produce the toxin themselves or whether bacteria on their skin do it for them. There is a wealth of studies begging to be done by motivated students of the future.

Appearance: Rough-skinned Newts are large salamanders with long tails that appear round when in terrestrial phase and paddle-like when in aquatic phase. They are generally maroon or black on top and orangish-yellow underneath. Their skin is typically bumpy, giving them the appearance of rough sandpaper, although while in aquatic phase their skin can become somewhat smooth. Single eggs are laid by attaching them to submerged vegetation, and they hatch into speckly larvae with dorsal fins and external gills to facilitate their fully aquatic lifestyle. Rough-skinned Newts overlap with other species of newts in some parts of their range. Distinguishing the species relies on examining their eyes. When the newt is viewed from above, the eyes do not bulge out past the line of the jaw.

Natural History: Rough-skinned Newts can be found in practically any coastal Northern California habitat, from forests to grasslands. They prefer to breed in ponds or slow-moving creeks, avoiding areas of rapidly flowing water. Like other newts in the genus *Taricha*, adults go through two phases over the course of a year: the aquatic phase and the terrestrial phase. While the aquatic phase is typically associated with mating, in some populations the newts will stay in the water for most of the year. Although patterns and timing vary dramatically among populations, in general Rough-skinned Newts emerge from terrestrial hiding places during winter rains and disperse to find ponds and mates. Their skin becomes smoother and their tails more paddle-like, and in spring the males develop darkened fingertips to help them grab onto females. And grab they do! It is common to encounter "mating balls" of newts, where multiple males clamor to grasp onto a single female. Once a female is ready to mate, the male deposits a sperm packet on the floor of the pond or stream, which she picks up with her cloaca. She then lays multiple single fertilized eggs (i.e., no clumps) attached to vegetation underwater. The larvae that hatch out remain aquatic for several months, breathing through the tuft-like gills extending from their cheeks. In the summer or fall, the larvae lose their gills and metamorphose into tiny, lung-breathing versions of adults. The metamorphs and adults leave the water, transition into the terrestrial phase, and travel overland to find a spot to spend the dry months. During this time newts can often be found hiding under logs or other cover objects. Adult newts eat arthropods, worms, and other invertebrates both on land and in the water, and larvae feed on tiny aquatic animals.

Range and Variations: Rough-skinned Newts occur from Santa Cruz County northward in coastal areas through Oregon, Washington, and Canada into southern Alaska, making them one of the most north-ranging amphibians in the world.

How to Find Rough-skinned Newts: Rough-skinned Newts can be found in ponds and other bodies of relatively still water in the early spring and summer (and, in some populations, well into the fall). Walk along the edges of ponds during the day and look for newts swimming slowly through the warm, shallow edges. They can also be found on land by flipping cover objects such as logs, rocks, or pieces of wood or tin. On rainy days, especially in the winter and early spring, newts can be encountered in large numbers on roads as they disperse to their mating ponds. If you choose to catch a newt, be aware that the toxin could potentially enter your bloodstream through a cut on your hand or if you handle food right afterward. This makes it all the more important to use gloves when examining amphibians.

Protection: none

RED-BELLIED NEWT

TARICHA RIVULARIS

FAMILY SALAMANDRIDAE

Although I have a lot more experience with California Newts (see page 114) because they live near me on the central coast, in my opinion the Red-bellied Newt is the most beautiful of the four newt species in California. Their bright red-orange undersides and black backs create such a stark and dramatic color combination. The Red-bellied Newt has the most restricted range of all four species. It is endemic to California, occurring only along a small coastal region north of the San Francisco Bay Area. Although I don't typically think of Sonoma, Mendocino, and other nearby counties as being centers of urban development, there is a lot of construction afoot in areas that were once open space. Due to heavy development for viticulture and housing, Red-bellied Newts are being impacted by habitat loss and road mortality. Fascinatingly, in 2009, a population of Red-bellied Newts was discovered in Santa Clara County, extremely far from their natural range. Although genetic analysis could not determine whether this population is natural or the result of human transport, it just so happens that the wayward population of Red-bellied Newts is not far from the lab of a Stanford University professor who conducted numerous studies in the 1950s examining whether species of *Taricha* could hybridize. This professor also happens to be the person who originally named and described the species in the 1930s. Although it could merely be an urban legend in the making, I cannot be tempted away from the notion that said professor or one of his students released the newts from their studies on purpose to establish a local colony or simply because they were not yet aware of the folly of doing so. We now know that animals should not be transported around the state and released because it could spread disease and disrupt normal genetic patterns. If you do catch a newt to admire it, or any other amphibian for that matter, avoid the temptation to keep it as a pet and instead release it at its site of capture.

Appearance: Red-bellied Newts are large salamanders with stark reddish-brown bellies and back coloration ranging from brown to black. They have completely black eyes, a feature that differentiates them from the other newt species in California. Eggs are laid under rocks in flattened masses, which hatch into aquatic larvae with speckled skin and fluffy external gills.

Natural History: Red-bellied Newts can be found in coastal forest habitats in the vicinity of streams and creeks. Like the

Spencer Riffle

Rough-skinned Newt (see page 105), the Red-bellied Newt alternates between a terrestrial phase and an aquatic phase associated mainly with mating. They appear to prefer flowing water and mate exclusively in streams. In the spring, males wait for females approaching the streams from the land and mob them, creating huge mating balls. Fertilization occurs when a female picks up a sperm packet left for her on the bottom of the creek by a courting male. She then lays eggs under rocks along the edges of the stream. The larvae remain aquatic for a few months, then metamorphose into little newtlings and venture onto the land, where they seem to disappear underground for several years. During these "lost years" of juvenile newt development, they probably feed and grow underground until they are large enough to emerge and return to the stream where they were born to mate for the first time. Adult and juvenile Red-bellied Newts eat many different kinds of invertebrates.

Range and Variations: Red-bellied Newts occur in Sonoma, Mendocino, Lake, and southern Humboldt Counties. There is also an isolated population in the Santa Cruz Mountains in Santa Clara County.

How to Find Red-bellied Newts: The best way to find Red-bellied Newts is to hike along streams in the spring, looking for the bright red bellies of newts within. In the late winter and early spring, you can also encounter them walking from their winter hideaways to their breeding creeks. These terrestrial-phase newts can be found by hiking through the forest or carefully driving along roads that transect forested habitat on wet days.

Protection: species of special concern (California)

California Newt. *Photograph by Max Roberts.*

SIERRA AND CALIFORNIA NEWTS

TARICHA SIERRAE AND T. TOROSA

FAMILY SALAMANDRIDAE

California and Sierra Newts are exemplary salamanders. What do I mean by "exemplary"? First, like other newts, they exemplify the term *amphibian* in the sense that they are a classic amphibious animal, spending part of the year on land and part on water. My students and I have been studying just how they are able to do this in collaboration with Haley Moniz, a postdoc in my lab. These newts' skin physiology changes when they are dry versus wet, possibly to help avoid desiccation during the long months of fall and winter until the rains beckon the newts back to their breeding ponds. It was easy to assemble a research team to conduct this study because my students love the newts and jumped at the chance to work with them. Which brings me to the next reason that newts are the consummate California amphibian.

Newts are exemplary in the sense of representing the best of their kind. What I mean by this is that most people consider newts to be cute and fascinating, so they act as good representatives of amphibians when it comes to public relations. I like to think of them as amphibian influencers: when people first encounter newts, the animals grab their attention, leaving them wanting more.

Newts are also exemplary in a different sense of that word. In addition to meaning "a classic example of something," *exemplary* can describe a deterrent that is used to ward off an attack. A low-key example of this is when a parent might ground one of their children over a relatively mild offense to set an example so that their other children don't act up. Newts advertise their stashes of highly poisonous tetrodotoxin with their bright coloration, which acts as an exemplary deterrent for predators. Indeed, hungry birds and other animals will often avoid this bitter, potentially deadly snack.

Finally, the word *exemplary* can mean serving as a model worth mimicking. Students of evolutionary biology know that toxic animals like newts often have mimics—animals that look like them but are not toxic. These mimics don't spend any energy making a poison but still enjoy the benefit of being avoided by predators fooled into thinking they are toxic. If you have read the other salamander species accounts in this book, do you have a guess as to who is mimicking newts? The answer is the Ensatina. It is no accident that the Yellow-eyed Ensatina subspecies, which co-occurs

The Yellow-eyed Ensatina subspecies (left) is thought to be a mimic of the toxic California Newts (right) that live in the same area.

with toxic newts along part of the California coast, not only has the same bright coloration as the newts but even has their yellow eyes! In a genius study, scientist Shawn Kuchta demonstrated that Western Scrub Jays avoid Yellow-eyed Ensatinas after encountering California Newts, whereas another Ensatina subspecies with less of a resemblance to newts did not enjoy this benefit.

California and Sierra Newts, along with other newts in this genus, can be very common where they occur and are easy to find and observe. They are great species for beginners to hunt. I hope you will head out to a wetland this winter or spring to observe the newts swimming around. You might get lucky and see a mating ball of males wrestling for access to a female, or a female laying eggs, or newts stalking aquatic invertebrates. Bring your kids, nieces and nephews, friends, and anyone else who needs a nature connection—newts are the ultimate amphibian kingpin, sure to grab their hearts and minds.

Appearance: California and Sierra Newts are similar in appearance, having been considered the same species until rather recently. They are large salamanders with rust-colored or brown backs and bright yellow-orange undersides. Their eyes are yellow with horizontal black pupils, and when viewed from above, the

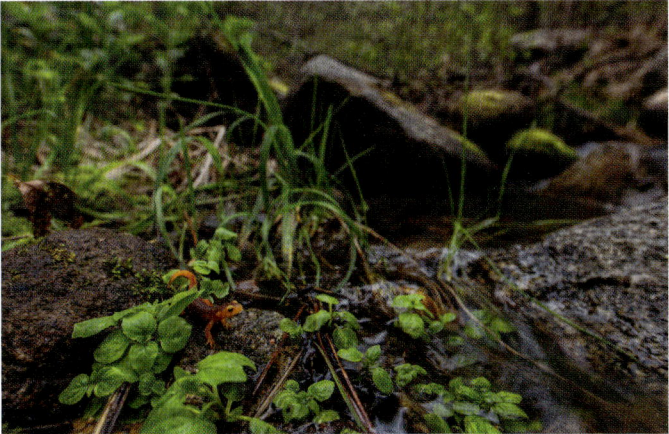

Sierra Newt. *Photograph by Marisa Ishimatsu.*

eyes bulge out past the line of the jaw. During their terrestrial phase, they have round tails and bumpy skin, and when they enter breeding ponds in spring, their tails slowly compress at the sides into a paddle shape and the skin becomes smooth, especially in males. Egg masses consist of multiple firm, gelatinous eggs attached in round masses to underwater vegetation or rocks. The larvae have external gills and are yellowish with jagged black bands on their back. Ensatinas are smaller as adults, have slightly more bulbous eyes, and have a constriction at the base of their tails that newts lack.

Natural History: California Newts (*Taricha torosa*) are commonly found in grasslands, chaparral, and many types of forests, whereas Sierra Newts (*Taricha sierrae*) are restricted to pine forests. Both California and Sierra Newts have a terrestrial phase occurring during the dry season and an aquatic phase stimulated by rainfall, though the timing of these phases depends on the local climate.

Generally speaking, winter and spring rains stimulate the newts to emerge from their dry-season hiding spots underground or underneath cover objects, then they migrate by crawling overland to their breeding wetlands. In some areas, mobs of newts can cover the ground, crossing roads by the thousands. California Newts breed in ponds or streams, and Sierra Newts prefer streams and can handle fast-moving water. During the aquatic phase, newts feed mainly on aquatic invertebrates. Mating is similar to other newt species in California. Males wait for females to enter the ponds then grasp them from behind, sometimes grappling with multiple other males for access to the females. A male eventually drops a sperm packet on the ground underwater, and the female picks it up with her cloaca. She then lays a few batches of eggs under the surface of the water attached to vegetation. After several weeks or months, depending on temperature, larvae hatch out and begin to hunt for tiny aquatic invertebrates. They typically metamorphose into newtlets in the fall.

Range and Variations: These newts are endemic to California; they live here and nowhere else. California Newts have a spotty distribution near the coast from the Bay Area south to San Diego County, and they also occupy the southern tip of the Sierra Nevada in Tulare and Kern Counties. Sierra Newts replace them in the Sierra Nevada from northern Tulare County north to Shasta County.

How to Find Sierra and California Newts: California and Sierra Newts can be found using the same methodologies we use to find other newt species. First, in parts of their range with mild climates, adult newts can be found by flipping cover objects in forests during the dry parts of the year, including fall and winter. Second, once winter rains commence, newts are out migrating

Larval California Newt. *Photograph by Spencer Riffle.*

overland both at night and during the day, and you can observe them by slowly driving through roads in their habitat. Be careful when driving to look for newts—thousands of them are run over each year, in some areas enough to cause populations to go extinct. Finally, perhaps the easiest way to find these newts is to walk around streams or ponds in the spring and look for newts swimming around. You might also see newt eggs by peering into these waters.

Protection:
Sierra Newt: none
California Newt: species of special concern (California)

California Chorus Frog.
Photograph by Max Roberts.

THE
FROGS

COASTAL TAILED FROG

ASCAPHUS TRUEI

FAMILY ASCAPHIDAE

Coastal Tailed Frogs are one of the Ten Wonders of California. Well, I don't think there is an official list, but if there were one, Coastal Tailed Frogs should be the singular California amphibian on it. Why? I can think of two main reasons. Which one you find more impressive might say something about your personality. First, two species of tailed frogs are the sole *extant* (meaning "still living, or not extinct") species of an ancient family of frogs called the Ascaphidae. Why ancient? Of the fifty-seven families of frogs and their relatives recognized worldwide, the Ascaphidae was the first to branch off from its ancient amphibian ancestor onto its own evolutionary trajectory, while all other extant frog families arose later. If you aren't impressed yet, listen up: this split happened around 200 million years ago, meaning that our very own diminutive Coastal Tailed Frogs are descendants of amphibians that began diversifying around the same time as the dinosaurs. This family therefore is different from other frogs in many ways, which leads me to the second main reason that Coastal Tailed Frogs are so noteworthy in California. They are the only amphibian in California, and the only frog *in the entire world*, to have a copulatory organ. Translation: they are the only frogs with penises. This is why they are called tailed frogs; the "tail," present only in males, is actually a copulatory organ that males use to transfer sperm to females. All other Californian amphibians have either external fertilization via spawning or internal fertilization via sperm packets picked up by females, but the Coastal Tailed Frog engages in lengthy bouts of sexual intercourse, sometimes "froggy-style" and sometimes "missionary" (face-to-face), sometimes for over twenty-four hours! If you are *still* not impressed, consider the fact that the "tail" can waggle around, controlled by a muscle that all other frogs lack. This same muscle is present in many mammals, however, and is the one your dog uses to wag its tail. As the saying goes, "Go home, Evolution, you're drunk."

Appearance: One of California's smallest frogs, adult Coastal Tailed Frogs max out at two inches in length. Their bodies are somewhat flatter than other frogs, reflecting their propensity to sidle into rock crevices in streambeds. The skin is typically covered in tiny bumps, and the color consists of mottled browns, greens, and dark reds to help them blend into the rocks and water. Coastal Tailed Frogs have black stripes crossing each eye that fade as they approach the snout, where they frame a pale, triangular nose tip. Unlike similar-sized Pacific Chorus Frogs, which have enlarged toe

Spencer Riffle

pads, Coastal Tailed Frogs have claw-like fingertips that they use for help to hold onto slick rocks. As described earlier, males have prominent "tails" that are actually penises. Eggs are laid in strings in rocky crevices underwater, and the tadpoles are dark brown or black with long tails tipped with white. If you are able to see the undersides of the tadpoles, they are immediately recognizable by their huge, sucker-like mouths.

Natural History: Coastal Tailed Frogs are associated exclusively with rocky streams in forests, where they live and mate. Occasionally they will stray within one hundred meters or so of these water bodies. They are mostly active at night. They use the rocks to hide from predators, anchor their eggs, and avoid being swept away by the water. Notably, if approached by predators or curious humans, the frogs will cannonball into the water and allow themselves to be carried downstream to safety. Coastal Tailed Frogs mate in the spring through the fall, with mating happening earlier in coastal populations than inland ones. The males do not call for females, presumably because their voices would be drowned out by the loud bubbling of the streams. Males and females find one another in some other (unknown) way, then males grasp females around their waists from behind. Though few observations of tailed frog mating have been made, it typically takes place underwater. The male inserts his copulatory organ and the frogs mate for an extended period, then she stores his sperm until she lays fertilized eggs in and under rocks in the stream in the summer. The eggs hatch within a few weeks into tadpoles that take at least one (but usually several) years to develop into adults. The mostly nocturnal tadpoles have fascinating, big, suction pad–like mouthparts for grasping onto rocks both to graze on algae and to prevent themselves from being swept downstream. Adults eat an array of invertebrates found in and near the streams.

Range and Variations: Coastal Tailed Frogs range from southern Mendocino County throughout the northwestern corner of California, then along the coast and Cascade Range north through Oregon, Washington, and British Columbia in Canada.

How to Find Coastal Tailed Frogs: Coastal Tailed Frogs can be found by walking along streams in good habitat with a flashlight at night during the spring, summer, and fall. The adults will often jump into the stream to evade you, so keep an eye out for them at a distance. Tadpoles can be seen foraging among the rocks at night. You can also flip streamside rocks during the day in the hopes of finding frogs hiding underneath.

Protection: species of special concern (California)

WESTERN TOAD

ANAXYRUS BOREAS

FAMILY BUFONIDAE

Some amphibians are prized by wildlife enthusiasts because they are rare or hard to find. Showing off photos of one of the uncommon Shasta salamanders or of a vanishingly rare wild Southern Mountain Yellow-legged Frog will get lots of oohs and aahs on social media. But there is also something to be said for loving the common species, the sure bets that you can reliably ogle when you need a little amphibian pick-me-up. The Western Toad is one such species. They are *everywhere*.

I can recall a zillion fond memories of these adorable little guys. When my husband and I bought our home, a huge Western Toad was there to greet us on our doorstep the very first night. While exploring the San Bernardino River in the middle of the

Mojave Desert in the spring of 2024, distressed at the number of invasive American Bullfrogs choking the tiny puddles in its sparse vegetation, I was relieved to see Western Toads hanging on in that most hostile of habitats. After a particularly rainy winter, I gawked at the sight of a huge, flooded lake literally packed to the gills (pun intended) with millions of tiny toad tadpoles, and returned a few weeks later to witness the shores teeming with metamorphs. When the once pristine swimming pool at a local ranch where I conduct research became neglected and effectively transformed into a dank, scummy pond, we placed pieces of wood on the surface to act as rafts for toads attracted to its waters, and ever since then we have been rewarded with jolly toads crowding these little lifeboats from spring through fall.

Western Toads are survivors. The stressors that have led to the demise of so many other California amphibians appear to slide right off Western Toads, for the most part. What makes them so resilient? For example, could they hold the key to pathogen resistance that might help us save endangered red-legged or yellow-legged frogs?

While we don't yet know the answer to that question and we don't even know yet what questions future scientists will ask about Western Toads and other resilient California amphibians, I know one thing for sure. The Western Toad is the perfect amphibian ambassador. Because it is common and adorable, the Western Toad has been cooed over, lovingly held, laughed at (but in a "laughing with" kind of way), and otherwise admired by thousands, probably millions, of Californians. We have a captive Western Toad named Cookie in my lab that we take to schools and kids' events, where children simply melt when they see her. (There actually have been several Cookies over the years, but don't tell anyone.) These chubby little toads have hopped their way into the hearts of many, and I hope that they now have a place in yours,

too. If you're not yet feeling the warm fuzzies about the Western Toad, go find them in the wild and report back. I already know what you will say (or, rather, coo).

Appearance: Western Toads are large, about the size of a fist. They are stocky with short legs, warty skin, and prominent oval glands behind each eye. Their color varies dramatically among populations and even individuals within populations, but they typically have large blotches and a thin light stripe extending down the center of their backs. Their eggs are laid in double-stranded strings and hatch into small, dark tadpoles.

Natural History: Western Toads can be found in practically any habitat in California. Although they are most common near water, they can be found hiding in rodent burrows or under cover objects away from water even during dry times of year. They can be active both day and night. In the spring, male Western Toads sit in the

water to wait for females. They have relatively quiet calls that they use mainly to argue with other males as they jostle for space, not so much to attract females. Males grab females from behind, then fertilize the strings of eggs as the females lay them. The eggs hatch within one to two weeks into small tadpoles that metamorphose into little toadlets in one to two months. Western Toads eat any small invertebrates that they can fit in their mouths.

Range and Variations: Western Toads live in all parts of California except for the eastern parts of the southeastern deserts. Two subspecies are recognized in California: the Boreal Toad (*Anaxyrus boreas boreas*), which is present in the extreme northern part of the state, and the California Toad (*Anaxyrus boreas halophilus*), which is present everywhere else. The Western Toad is also widely distributed outside of California, ranging as far north as southern Alaska, east into parts of Wyoming, Colorado, and New Mexico, and south into northern Baja California, Mexico.

How to Find Western Toads: You could run into Western Toads in many places and situations. The best way to look for them is near ponds and streams during the peak mating season, which is mainly spring but can occur in winter in warmer areas and in early summer at high elevations. Look for toads sitting along the water's edge or in the water with their heads poking out. In addition, Western Toads can be found by flipping cover objects like logs, flat rocks, cover boards, and junk. As temperatures rise and the habitat dries out, toads often move deep into rodent burrows until it rains in the winter, but sometimes they find moist microhabitats to congregate. I can find Western Toads any time of year in a shed in my yard, where our irrigation system drips a small amount of water during the spring and summer. Creating amphibian-friendly habitats in your yard (see page 22) is likely to attract Western Toads and Pacific Chorus Frogs as the main residents.

Protection: none

Ryan Sikola

ARROYO TOAD

ANAXYRUS CALIFORNICUS

Arroyo Toads are a sad chapter in California's amphibian history. On the surface, they are a federally endangered species that we are witnessing being snuffed out right before our eyes, largely due to the actions of our own species. However, the full narrative is more complex. Arroyo Toads are habitat specialists, stubbornly insisting on occupying habitat that is rapidly going extinct in California. As a result, so are they.

These toads occupy sandy streams that hold enough water to resist drying out before the toads can complete their reproductive cycle. Such streams are now rare in California. They were a lot more common in the distant past up until about ten thousand years ago, when California's rainfall declined drastically and its habitats subsequently were remodeled into a much more arid landscape. Arroyo Toad populations started to become fragmented, remaining only in those sandy drainages with the appropriate geology and geography to hold enough water for them. More recently, multiple stressors caused by humans have hit the remaining populations of this species very hard. Destruction of their rare stream habitat by off-road vehicles, mines, hikers, and cattle, plus infection by chytrid fungus and other pathogens, predation by invasive bullfrogs, crayfish, and fishes, and stress from recent major droughts have all come together to push the Arroyo Toad onto a precarious perch as one of California's most endangered species. Luckily, this little toad is rather tough, stalwartly clinging to existence, especially in coastal populations. A recent study resurveyed Arroyo Toad populations after twenty years of drought conditions and found that they had disappeared from some sites but were still present at most. Scientists are hard at work developing plans to protect sensitive populations by extirpating invasive species, restricting human recreation, and diverting water to flow into their drainages. Note that this restricted recreation also applies to people searching for this toad. In this account, you will learn about this hardy little amphibian, but I urge you not to go searching for it in the wild. Leave our sensitive toad species alone and instead satisfy your needs to find and hold adorable toads with the widespread and very successful Western Toad (see page 126).

Appearance: Arroyo Toads are small to medium-sized toads, usually mottled with gray and brown colors. They have warty skin, lack stripes down their backs, have oval glands behind each eye, which are lighter in front than in back, and usually have a pale,

Ryan Sikola

boomerang-shaped stripe across their eyelids and heads. Their bellies are pure white in color. Their eggs are laid in long double strands, often twisted into a DNA-like helix shape. The tadpoles are very dark, and as they grow they begin to express a mottled pattern that helps them blend in with the sandy stream banks.

Natural History: Arroyo Toads are one of the strictest habitat specialists of all California amphibians. They occupy sandy streams with small pools along the banks, associated with vegetation like oak trees, willows, and cottonwoods. They are mainly nocturnal. Male toads call for females from these small pools in the spring, luring them in with their extended trilling vocalizations. When a female chooses a male, he grasps her around the back by the armpits, and she lays two long strands of eggs in the shallow pool along the stream bank as he fertilizes them. The eggs develop in a few inches of water, not attached to vegetation. Tadpoles hatch within a week. They are poor swimmers for a few days but are

protected from predators by toxins that the mother deposits into the eggs. However, stream flooding or other disturbances can wash eggs and tadpoles away, which is just one of many reasons this sensitive species is endangered. The tadpoles filter feed along the sandy stream banks, while the adults gorge themselves on night-active ants. Arroyo Toad tadpoles metamorphose onto the stream banks within a few months, typically in the summer. Unlike the more nocturnal adults, the toadlets sit in the wet, sandy banks in full sun, eating flies and ants and avoiding dehydration by absorbing water across their lower bellies.

Range and Variations: Arroyo Toads live in isolated populations from Monterey County southward into northern Baja California, Mexico. Mostly they are found in coastal counties, with a couple of populations extending into Mojave Desert areas of Los Angeles and San Bernardino Counties.

How to Find Arroyo Toads: Much of the recent decline in Arroyo Toads was caused by destruction of habitat by humans. It is our responsibility to ensure that we don't cause further destruction, which could easily happen if we accidentally crush them with our vehicles or feet, or introduce pathogens via dirty boots. Those bent on finding Arroyo Toads can use the natural history information in this account or in a field guide to learn how to identify good spots to search, listen for males' calls, and look for nocturnal adults or diurnal toadlets. If you do observe an Arroyo Toad, please post it to iNaturalist so that experts can review your identification and learn more about the species' status. A pro tip is that you can (and should) obscure the location when posting sensitive species like Arroyo Toads to protect it from people who might visit and destroy habitat; the scientists can still access the location data, but others cannot.

Protection: species of special concern (California), endangered (federal)

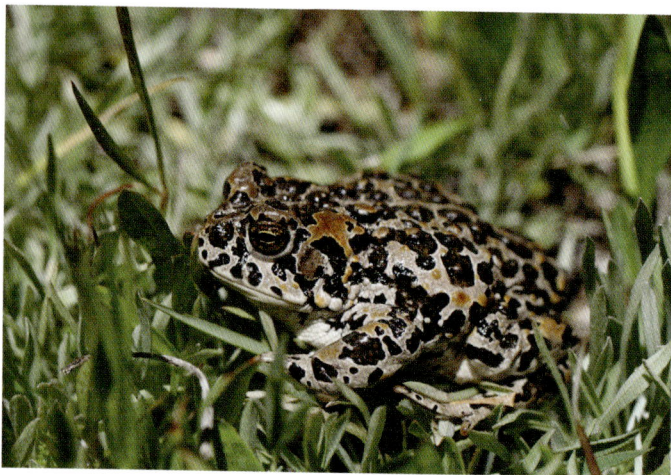

Female Yosemite Toad. *Photograph by Ralph Cutter.*

YOSEMITE TOAD

ANAXYRUS CANORUS

Endemic only to the Sierra Nevada and gone from half of its historic range, the Yosemite Toad was declared threatened in 2014 by the US Fish and Wildlife Service, and the state of California protects it as a species of special concern. What makes the case of the Yosemite Toad particularly worrisome is the fact that they live mainly in very remote areas, far from human development, in the high-elevation reaches of the Sierra Nevada. Heck, they are named after the pristine national park in which they live. Yet, even this toad has been disappearing. Why? It is difficult to disentangle the complex interactions among the stressors that may be impacting the Yosemite Toad. They have experienced major die-offs in the past from chytrid fungus and severe drought, and other factors like increasing ultraviolet radiation in their high-elevation habitats could play roles as well. The toads living inside Yosemite National Park have a huge support team, though. One of my former students is a herpetologist for the National Park Service. We met up on a recent visit to the park, and she told me that park biologists have reintroduced Yosemite Toads to some areas of the park from which they had disappeared (notably, the biologists also successfully released California Red-legged Frogs and Sierra Nevada Yellow-legged Frogs after mounting an aggressive campaign to rid the park of invasive American Bullfrogs). I loved hearing this news because it shows how incredibly important our parks are to the survival of this beautiful species of toad. I am so proud that one of my students is on the front lines of the rehabilitation of the Yosemite Toad. Take my tax dollars and give it to the toads!

Appearance: Yosemite Toads have an unusual distinction when it comes to their appearance: the males and females look really different. Females are about the size of a small fist and have a very distinct dark, blotchy pattern on a tan or yellowish background, and males are smaller and have a more solid green color. They are warty, have an oval-shaped gland behind each eye, and typically lack a stripe down the back. Eggs are laid in strands, tadpoles are dark in color, and metamorphs have a stark, blotchy coloration resembling that of adult females.

Natural History: Yosemite Toads inhabit wetlands and willow forests in mountain meadows of the Sierra Nevada. They hide underground throughout the cold winters that characterize their habitat, then emerge in the late spring to reproduce. The toads endearingly crawl on their tiptoes across the snowy meadows to

Male Yosemite Toad. *Photograph by Jackson Shedd.*

enter the pools of snowmelt. Males call for females from shallow ponds or meandering streams with a loud, high-pitched, continuous "wa-wa-wa-wa-wa" sound that makes me think of the alarm clock that terrorized my mornings during high school. A male grasps a female's armpits from behind until she lays strings of eggs that he fertilizes. These eggs hatch within a couple of weeks into small, dark tadpoles, which metamorphose later that summer or fall. Yosemite Toad tadpoles eat detritus and algae, and adults eat an assortment of invertebrate prey, mostly in the meadows and forest edges during the summer after they leave the breeding ponds.

Range and Variations: The Yosemite Toad is endemic to California, inhabiting only high elevations in the central Sierra Nevada.

How to Find Yosemite Toads: Because they occur at high elevations, Yosemite Toads exhibit highly seasonal activity. The best time to look for them runs from April through September, with midsummer being best at very high elevations. To find Yosemite Toads, walk around the edges of wetlands and in the surrounding meadows looking for the toads active during the day. It's not uncommon in some parts of Yosemite National Park to see a toad hopping across a trail. You can also listen for the males calling in the evening (and sometimes even during the day). Be sure not to touch, capture, or harass the toads, and you should always clean your footwear in a mild bleach solution before walking in or along the edges of wetlands to prevent accidentally transmitting pathogens into the water.

Protection: species of special concern (California), threatened (federal)

Yosemite Toads in amplexus in a wetland filled with strings of eggs from previous matings. *Photograph by Rob Grasso.*

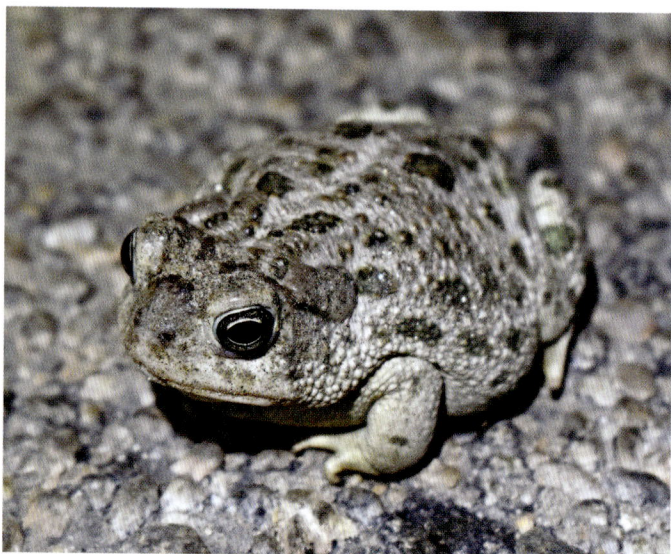

GREAT PLAINS TOAD

ANAXYRUS COGNATUS

FAMILY BUFONIDAE

The mighty Great Plains Toad always conjures images of amphibian hunting in New Mexico for me. One night about fifteen years ago, I was night driving through the southwestern Bootheel in pursuit of the trailing ends of monsoon storms, which tend to bring out the critters. We kept the car radio off and the windows down to listen for hollering amphibians breeding in the recent storm waters. As we drove through a flooded embankment on a remote desert road, our ears told us it was time to stop. The water was absolutely loaded with Great Plains Toads, the males calling their little hearts out and groping around for females to grab onto. The sound was so deafening that we literally had to scream at the top of our lungs to one another even when standing side by side! Great Plains Toads are one of the most common amphibians in— you guessed it—the American Great Plains and southwestern deserts, but in California they are much more restricted. They were originally found only in the extreme southeastern part of the state and now can also be found in the Imperial and Coachella Valleys, where agricultural development has likely provided enough water resources to allow them to spread.

Appearance: Great Plains Toads are large (approximately fist sized) toads with the traditionally "toady" wart-covered skin and elongate glands behind their eyes. They are usually light in color, with symmetrical green blotches on their sides separated by a light stripe going down the center of their backs. Eggs are laid

Sean Barefield

The vocal sac of a male Great Plains Toad inflates when he is calling for mates.
Photograph by Jeff Martineau.

in long strings. Tadpoles are small and black, showing a mottled appearance shortly before metamorphosing into small toadlets.

Natural History: Great Plains Toads occupy grassland and desert habitats and also do well in disturbed areas with plenty of water resources, such as agricultural lands. These toads hide underground in the hot summer, winter, and other times when water is not available, then emerge during spring and summer rains to breed in ponds, temporary pools of rainwater, or flooded agricultural fields. Have you ever found yourself in a room with a loud cricket? If so, you have some idea of what the Great Plains Toad's call sounds like—a continuous, sped-up cricket yell. When a female chooses a male, he sits on her back and grabs her around her armpits until she lays eggs in long strings, which he fertilizes. Adults feed on insects and other arthropods, while tadpoles graze on algae in the water.

Range and Variations: Great Plains Toads have a large geographic range from southern Canada through the American Great Plains into northern Mexico. In California, they can be found in the Coachella and Imperial Valleys and along the Colorado River.

How to Find Great Plains Toads: In California, Great Plains Toads can most easily be found by searching for breeding toads at night near agricultural areas. When farmers flood fields in the winter and spring, or after heavy rainfall, toads emerge to mate in almost any body of water, provided the water isn't moving too fast. The easiest way to find them is to listen for their calls, then use your flashlight to find them sitting in the water.

Protection: none

Ceal Klinger

BLACK TOAD

ANAXYRUS EXSUL

~~~~~~

FAMILY BUFONIDAE

In a remote corner of California's eastern desert lies a dusty desert valley with something very special: year-round artesian springs. This means that the water is forced up from deep underground. There's nothing so delicious as a permanent source of water in the middle of the desert, especially when it comes to amphibians. The springs in Deep Springs Valley in Inyo County are occupied by a fascinating California endemic species, the Black Toad, that occurs naturally there and nowhere else. Relatively recently in geologic time, more than half the valley was covered in a lake. But now the amphibian habitat consists only of a series of cold springs that edge the valley. Because of its tiny range, the Black Toad is protected by the California Department of Fish and Wildlife. The Wikipedia entry for Deep Springs Valley references the Black Toad with the following statement: "The population of toads, however, appears to be stable, despite the presence of cattle that graze around the lake during the winter and spring." Faculty, staff, and students from the nearby Deep Springs College, a unique institution emphasizing labor and self-governance, indeed run cattle in the valley as part of their program. On the surface, this might raise your eyebrows, as grazing cattle in a desert environment is not typically considered an environmentally friendly practice. However, the cattle are managed carefully, including blocking off their access to parts of the springs and running them in the area mainly in the winter, when the toads are dormant. The cows feed on emergent vegetation, keeping the springs habitat open enough for toads to flourish when they emerge later in the year. The cows also leave behind cow pies that attract insects, which the toads feed upon. By properly managing the cattle, the college effectively uses them to replace the ecological function of the native grazers like Bighorn Sheep that once roamed the land. This is a delicate balance, but to date the Black Toad seems to be doing fairly well.

Appearance: Black Toads are—you guessed it—black, with a small amount of whitish mottling plus a white stripe down the center of the back. These are medium-sized toads, with warty skin and oval glands behind the eyes. The eggs are laid in strings in shallow water, and they hatch into tiny, dark tadpoles that metamorphose into dark green toadlets.

Natural History: Black Toads are found exclusively in and around a series of desert springs in a tiny region of California (see Range and Variations). Their activity follows the temperature: in the

Strings of silt-covered eggs surround a pair of mating Black Toads.
Photograph by Ceal Klingler.

spring and fall when temperatures are mild, they can be found during the day, while in summer they are active mainly at night. They retreat underground to hibernate during the colder months from fall through early spring, when they emerge to reproduce. Mating behavior in Black Toads is somewhat unusual. Males don't call for females, nor do they throw themselves at passing females, like many of the more eager toad species do. Rather, they typically call to communicate with other males as they spar for space in shallow water during the mating season. Their calls remind me of my backyard chickens clucking or a gaggle of geese flying overhead in the distance. When a female approaches a male, he grasps her around the armpits from behind, and she lays long strings of eggs among underwater vegetation as he fertilizes them. These hatch into small black tadpoles that meta-morphose later that summer. Adult Black Toads eat invertebrates in and around the springs, and the larvae graze on algae and other organic material in the water.

Range and Variations: The only natural population of Black Toads is in the Deep Springs Valley in Inyo County, making it have one of the tiniest geographic ranges of any amphibian in the world.

How to Find Black Toads: Think twice before going on a pilgrimage to search for Black Toads. Due to their protected status, you are prohibited from even touching a toad. Also, much of their habitat lies on private property owned by Deep Springs College. If you do go looking for them, you can try walking around the edges of springs during the day in the spring or fall, or at night in the summer, watching for dark toads sitting in the shallow water. It is imperative that you bleach your footwear (see page 22) to avoid accidentally transmitting pathogens into the waters.

Protection: threatened (California)

Max Roberts

RED-SPOTTED TOAD

ANAXYRUS PUNCTATUS

If you are looking to find amphibians in the driest parts of California, the Red-spotted Toad is your critter. It was first discovered in Death Valley, one of the hottest and driest places in the world. In the Mojave Desert, it rains fewer than twenty days per year, and many of those are in winter, when the amphibians are hibernating underground and cannot fully access the water. It is impressive for any animal to survive such conditions, let alone an amphibian whose biology is partly defined by their leaky skin. And the Red-spotted Toad is not just surviving. It is extremely common and showing very few signs of the declines plaguing so many other amphibians in California. So, how *does* the Red-spotted Toad survive and thrive in the desert?

The Red-spotted Toad has been a model for studies of amphibians living at the edges of their thermal and hydric tolerances. All life stages of this species can survive higher temperatures than other species of toads, as one might expect due to their adaptations to living in deserts. In fact, one study showed that some aspects of their immune response to pathogens work best at extremely high temperatures. Red-spotted Toads burrow underground and remain inactive during dry times of year, and when it rains they pop out en masse to engage in a process with a name I find incredibly endearing: cutaneous drinking. Their lower belly, or "pelvic patch," has special water channels. When they come across wet sand, they squat down and dip their water channel–rich pelvises into the substrate to literally drink across their skin. I like to picture a cocktail bar where the toads sidle up and tell the bartender what they feel like that day, and then instead of a beverage the bartender mixes up custom concoctions for the toads to sit in and absorb. If that's not a fun enough image for you, consider the fact that scientists have discovered that Red-spotted Toads use sensors on their feet to detect the composition of the water they encounter. So, we can add into our little bar scene an image of the customers dipping their hands into a proffered concoction, finding it too "wet," and asking their bartender to add some salt to the rim. Now that I am craving a margarita, let's take a look at the biology of this diminutive desert denizen.

Appearance: Red-spotted Toads are small toads (two to three adults could fit in your palm) with round glands behind the eyes, and they are grayish-brown with tiny warts. These warts are usually red or orange, giving the toad its name. Interestingly, Red-spotted Toads are unique among California toads in not laying strands of eggs; instead, the eggs are laid singly, often sticking to others to form masses. The tadpoles are small and dark, often forming dense aggregations in small ponds or puddles as they dry up.

Natural History: Red-spotted Toads inhabit wetlands, especially rocky streams, in California's southeastern deserts and some associated grasslands and woodlands. They can also appear in temporary ponds like roadside ditches following heavy rains in the spring and summer, when they emerge from their underground hideaways to mate and feed. A male Red-spotted Toads calls from the water or along its banks with a high-pitched, trilling sound, then grasps a female by the armpits until she lays eggs one at a time in the water while he fertilizes them. Tiny, dark tadpoles emerge from the eggs within just a few days, then metamorphose into similarly tiny toadlets that crawl onto land and crowd the shores of the pond for a while. Red-spotted Toads, like most Californian anurans, eat insects and other small invertebrates.

Range and Variations: In California, the Red-spotted Toad ranges through the state's southeastern deserts from the vicinity of Death Valley south through all of Baja California and large parts of Sonora, Mexico. They also occur in the American deserts as far east as Oklahoma and Texas.

How to Find Red-spotted Toads: Red-spotted Toads can be present in large numbers in certain areas in the desert, especially around streams, small human-made water bodies, and temporary pools that fill with rainwater. Walk around these areas in the

dark hours following sunset, using a flashlight or headlamp to spot small toads hopping about. After rainstorms, you can often encounter Red-spotted Toads while driving on roads transecting desert habitat, so be careful when night driving during or after storms. From your vehicle, these amphibians appear as light-colored, toad-shaped rocks in your headlights. Finally, Red-spotted Toads can often be found in some neighborhoods in desert cities, hiding in wet refugia like irrigation sheds, and may be found by flipping cover objects in backyards and appropriate desert habitat.

Protection: none

Spencer Riffle

WOODHOUSE'S TOAD

ANAXYRUS WOODHOUSII

Most of my experience with Woodhouse's Toads has been outside of California. In 1997, I spent the summer between my junior and senior years of college as a field technician doing amphibian and reptile surveys in South Dakota. I swam around cattle impoundments catching garter snakes by day, and drove around the grasslands listening for frog calls by night. The soundtrack of that summer was half Johnny Cash and half toad calls. Let me tell you, the din made by hollering toads at night in the northern Plains is impressive. At many of our sites, the bleating calls of Woodhouse's Toads harmonized with the trills of Great Plains Toads, burning the sounds of these two toad species into my brain forever.

Woodhouse's Toads could be described as "California adjacent." They are widespread and common throughout the central part of the United States, with their native range just barely popping into extreme southeastern California along the banks of the Colorado River. However, the Woodhouse's Toad has been steadily marching and swimming its way into the irrigation canals of the Imperial and Coachella Valleys over the past century or so, greatly aided and abetted by people. In the early twentieth century, major floods in the Colorado River likely whooshed some toads into the Imperial Valley, and they have been slowly spreading from there ever since. This area of California would normally be too arid for Woodhouse's Toads to thrive, but expansive agriculture has created a network of waterways that is helping these toads disperse.

Appearance: Woodhouse's Toads are large (up to the size of a fist), with warty, blotchy skin in shades of tan, green, and brown. Adults have a pale stripe running down their backs. The glands behind the eyes are elongated, more so than the glands of other toads found in the same area, like Great Plains Toads. Eggs are laid in long strings. Tadpoles appear small and dark, and on closer inspection they are covered in little metallic splotches, with tails that have pale undersides.

Natural History: Woodhouse's Toads in California live mainly in irrigation ponds, canals, ditches, water hazards on golf courses, and impoundments in agricultural areas, and they also occur along the lower Colorado River. They are nocturnal and can be found active from early spring though early fall. Males gather at night in the water, making loud calls similar to the sound a sheep would make if it bleated and snored at the same time. Males grasp the larger females from behind by the armpits, and the females lay

strings of thousands of eggs as the males fertilize them, where the eggs attach to underwater vegetation. Tadpoles are often found in high densities in the water, where they feed on detritus and algae for up to two months before metamorphosing into little toadlets. Adults feed on most any invertebrate that they can find.

Range and Variations: As described on the previous page, Woodhouse's Toads in California are native only to extreme southeastern California along the Colorado River and have extended into the Imperial and Coachella Valleys of Imperial and Riverside Counties within the past century or so. Outside of the state, Woodhouse's Toads occur in a wide stripe through the center of the United States, from northern Montana and North Dakota down to northern Mexico.

How to Find Woodhouse's Toads: Woodhouse's Toads can be extremely common in some areas, especially in the heart of their range in the Plains states. In California, the best way to find them is to drive around at night on roads near canals, farms, and irrigation ditches in the Imperial and Coachella Valleys. Keep your windows down and the radio off to listen for the calls of toads in the distance to help guide you to a good spot. Mating can occur anytime in the spring through early fall, especially on rainy nights. You can also watch for them crossing roads in their peculiar "run-hopping" style as you drive.

Protection: none

Dustin Smith

COMMON COQUÍ

ELEUTHERODACTYLUS COQUI

The Common Coquí (pronounced "ko-KEE") is native to Puerto Rico, but invasive populations have become established in many other places, possibly including California. By "possibly," I mean that individuals of this frog have been found in multiple places in Southern California, but it is not 100 percent clear that sustaining populations have become established. This tiny frog has spread by hitchhiking in vegetation, mainly tropical plants shipped around the country via the nursery trade. It requires warm, humid conditions like in its native Puerto Rico, and therefore could become established in areas with similar climates. In suburban Southern California, there are places that are warm enough, and irrigation

A male Common Coquí calls for females.
Photograph by Dustin Smith.

of lawns could provide enough moisture. Chances are that you will never encounter this frog, but I chose to include it in this guide because biologists in California are very keen on preventing it from becoming established or spreading further, so in the unlikely event that you do come across one, I want you to be equipped to identify it and report it online on iNaturalist and via the California Department of Fish and Wildlife's Invasive Species Program.

Appearance: Common Coquís are tiny frogs with unwebbed feet and round toes that allow them to climb around on vegetation and on structures. They come in many colors, often brown or gray, usually with a thin white stripe down the center of their backs.

Natural History: This is a tiny frog with a huge voice. Males call from their nighttime perches among vegetation with an unbelievably loud "ko-KEE!" The few individuals found in California have been living among tropical plants. These frogs do not require standing bodies of water for breeding. A female approaches a small calling male, he climbs onto her back and grasps her, and she lays eggs on plant leaves or small crevices in the ground where water has accumulated. The male fertilizes the eggs as she lays them, then he stays with the eggs and guards them until they hatch into tiny froglets. Common Coquís eat a wide array of small insects and other invertebrates.

Range and Variations: In California, Common Coquís have so far been reported and confirmed from several homes, plant nurseries, and beach areas in San Diego, Orange, and Los Angeles Counties. Though native to Puerto Rico, these frogs have established populations in other areas including South Florida, Hawaii, and several Caribbean islands.

How to Find Common Coquís: As stated earlier, you are unlikely to find Common Coquís in California. If you go on vacation to Puerto Rico or Hawaii, you are most likely to take note of their presence at night from their deafening yells. I once visited the Caribbean island of Aruba, where our resort was crawling with invasive Common Coquis; the resort provided earplugs to sleep at night! You can follow the sound of the calls and, with patience, uncover the tiny males responsible, usually sitting within vegetation.

Protection: none (non-native)

Spencer Riffle

CALIFORNIA CHORUS FROG

PSEUDACRIS CADAVERINA

FAMILY HYLIDAE

The California Chorus Frog, sometimes called the California Treefrog, may be the most common Southern Californian amphibian you have never heard of. One reason for this is that this species is so closely associated with streams that you will never find them more than a few short steps away from the water. But if you explore most any boulder-strewn creeks and streams within their range, you could see huge numbers of them. You *could*, that is, if you were actively searching for them because you read about them in this book and know how to look for them. Otherwise, you might walk right by a frog, even dozens of them, without noticing. Why?

A "choir" of California Chorus Frogs huddle together, probably to help conserve water. *Photograph by Max Roberts.*

California Chorus Frogs are masters of disguise. They rely on camouflage to help them blend into the boulders they perch upon so that predators don't see them. The rocky canyons they inhabit can be really hot, even in the evening, when they are most active, so they rely on the evaporation of water across their skin to keep them cool even when temperatures are sweltering. This means they need to be able to replenish that water by soaking it up across their skin, hence their consistent association with water during their active season. They also have an adorable habit of cuddling up in big clusters of multiple frogs, probably to help reduce water loss across their skin at times when they don't need it for thermoregulation. Tadpoles and metamorphs also aggregate in large groups, and while we could speculate on adaptive reasons for this, it is nice to simply acknowledge that California Chorus Frogs are social at each of their life stages. Perhaps they just like being together?

You know how people have made up cute names for groups of animals, like a skulk of foxes, a murder of crows, and a rhumba of rattlesnakes? Well, the established name for a group of anurans is an army of frogs. While this name could work well to describe the huge numbers of toads marching overland to and from a mating pond, it just doesn't sit well for me when it comes to California Chorus Frogs. They need a more peaceful group name that better describes their gregariousness and performative spirit. I therefore propose this as a name for a group of California Chorus Frogs: a choir of chorus frogs. Go out this spring and find a choir, and report back: can you think of a better group name? I will gladly entertain suggestions!

Appearance: California Chorus Frogs are medium-sized frogs with the enlarged toe pads that are characteristic of all

Max Roberts

treefrogs. They have small blotches on a background of gray or tan, typically matching the rocks and other substrates where they live, with light bellies. The eggs are in small clusters attached to underwater vegetation in areas of streams with slow-moving water and plenty of rocks. Tadpoles also have little blotches on their backs and light undersides, and their bodies are somewhat flattened top to bottom.

Natural History: California Chorus Frogs live almost exclusively in and around creeks and streams. In the spring and summer, they wedge themselves into crevices near the water during the day, then in the late afternoon they leave their shelters and move to the edge of the water, often on rocks, where they hunt for insects and males call for mates and tussle with other males. They stay there until about midnight, when they head back to their crevices until the following afternoon. In the fall, they climb further up the banks of the stream, where they take refuge inside deep, moist crevices for the inactive winter season. The mating call of male California Chorus Frogs sounds like a brief, high-pitched quack made over and over. In the water, a male grabs a female around her waist, then the female lays eggs one at a time while the male fertilizes them. While this can occur anytime from late winter through fall, most mating happens in the spring. The eggs typically hatch within a month, then the tadpoles graze on aquatic vegetation for a month or two until they metamorphose into froglets that tend to gather around the water's edge for a time.

Range and Variations: California Chorus Frogs occur in a stripe of near coastal habitat ranging from San Luis Obispo County south into northern Baja California, Mexico.

How to Find California Chorus Frogs: You can find California Chorus Frogs by hiking around rocky streams in the late afternoon or the hours following sunset, using a flashlight to locate frogs sitting on rocks within a few feet of the water. You can also listen for calling males to help you find them. During the daytime, your best bet is to peer into rock crevices alongside the water, where you might see California Chorus Frogs crammed inside to avoid the dryness and heat of the day.

Protection: none

Max Roberts

PACIFIC CHORUS FROG

PSEUDACRIS REGILLA

~~~~~~

FAMILY HYLIDAE

There is a huge party going on right now down the street from my house. I can hear the noise from indoors, even with the windows closed. What is crazy is that it happens almost every single night in January and February. As soon as the sun goes down, the party-goers arrive and start reveling, on some nights not stopping until dawn. But I don't plan to call the cops to come break up this party.

That is because the rowdy partiers are Pacific Chorus Frogs, also commonly known as Pacific Treefrogs. Every winter, the first big rainstorms flood a field in my neighborhood, setting the stage for the Coachella of frog festivals. Thousands of tiny Pacific Chorus Frogs emerge from their refugia underground, under the bark of trees, or under woodpiles or hot tubs in my neighbors' yards, and hop on over to the party grounds. The males find their spots and start singing their little hearts out in the hopes of finding dates for the night. The females strut around, listening to the chorus of these frogs and comparing the posturing males until they find one they like. Then the party turns into an orgy of the most wholesome kind: thousands of frogs sit in the water, paired up froggy-style, until that special moment in which a female lays her eggs and the male ejaculates onto them. This is anuran romance at its finest, and I am so lucky to host parties like this in my neighborhood.

You could be just as lucky if you go outside at night, tune out background noise, and listen. Pacific Chorus Frogs are practically ubiquitous in California, except for the southeastern deserts. Parties like the one in my neighborhood take place all over the state, with the frogs' song forming a major component of California's soundtrack. I mean this literally—sometimes when watching movies that are supposed to take place in some faraway locale, you will hear the distinctive call of a Pacific Chorus Frog, a dead giveaway that the movie was actually filmed on a set in Los Angeles or that it used prerecorded Pacific Chorus Frog calls as nighttime background soundtracks. The call of the Pacific Chorus Frog is so ingrained as a standard frog sound in people's minds that filmmakers use it even if the movie takes place in Africa because apparently using the calls of frogs native to the film setting would confuse American movie watchers!

Pacific Chorus Frogs are certainly the most widespread and conspicuous amphibian species in California. If the official state amphibian of California was determined based on ubiquity, it would undoubtedly be the Pacific Chorus Frog. However, that honor was bestowed on the California Red-legged Frog (see page 175), probably to draw attention to the plight of this threatened

species. Pacific Chorus Frogs, on the other hand, are anything but threatened. They are extremely common in habitats ranging from coastal scrub to more than 11,000 feet high in the Sierra Nevada, and are notable in that they commonly appear in people's yards. Due to their sheer numbers, these frogs play important roles in regulating insect populations, acting as a food resource for numerous bird and reptile predators, and helping to traffic nutrients between terrestrial and freshwater habitats. But just because Pacific Chorus Frogs are not threatened doesn't mean that they don't succumb to the myriad of stressors that impact many of our other California amphibians. They do, and their absence in any given area is used as an indicator of poor environmental health. In this way, the Pacific Chorus Frog is an exemplary canary in a coal mine, as discussed in the front matter of this book.

Given all the important jobs done by Pacific Chorus Frogs, it's the least we can do to let them party loudly outside our windows all night long.

The vocal sac of a male Pacific Chorus Frog amplifies his calls to attract females.

Spencer Riffle

Appearance: Pacific Chorus Frogs are small treefrogs with enlarged toe pads, smooth skin, and a distinctive black Zorro-style mask extending across the nose and eyes to the shoulders. They are most commonly green or brown, often with dark blotches, but their background colors can be more variable. While this coloration is genetically determined, the frogs can also change color fairly rapidly based on environmental variables like temperature. The underside of the body is light in color, often with a yellow hue on the legs. Egg masses are small clusters of several dozen eggs laid under the surface of the water attached to plants. Tadpoles are dark on top and light below, and have an iridescent sheen to them.

Natural History: Pacific Chorus Frogs can be found in just about any habitat in California. They breed in streams, ponds, lakes, and temporary water bodies like the field in my neighborhood, and they are often observed far away from water during the non-mating season. These frogs can be active day and night, though in hot areas they tend to hide during the day, and in cold areas they hibernate for the winter. Typically stimulated by winter rains, Pacific Chorus Frogs gather in the water where males holler for females with a series of call types, most commonly the classic "ribbit." Males grasp females around the necks from behind, then fertilize the eggs as they are laid. The eggs hatch within a few weeks, then the tadpoles feed on tiny aquatic critters and decaying material until they metamorphose in a couple of months. Metamorph and adult Pacific Chorus Frogs feed on small invertebrates.

Range and Variations: The Pacific Chorus Frog occupies all of California except for much of the southeastern deserts, and even there they can be found in some oases and seeps. A 2006 paper by Ernesto Recuero and colleagues split the frog into three species, but most scientists no longer recognize these species because the frogs seem very similar to one another in most ways.

How to Find Pacific Chorus Frogs: Pacific Chorus Frogs are very easy to find. In the winter rainy season, go outside at dusk with a good flashlight and listen for their calls. When you approach a calling male, he will often stop hollering for a while, making it challenging to find the tiny frog among the aquatic vegetation. With patience, you will find him! Pacific Chorus Frogs are great frogs to watch. If you find a spot where males are calling and you sit down and remain motionless, they will often restart their calling, and you might even see males in territorial spats or a mating pair creating the next generation of Pacific Chorus Frogs. It is easy to find and watch their tadpoles, as these are the most commonly seen tadpoles in California. You can also find adult frogs by flipping cover objects, especially in areas with moist soil. If you are intrigued by the idea I presented in the front matter of this book of creating amphibian-friendly habitat in your yard, a great place to start is targeting the Pacific Chorus Frog. You could go all out and make an artificial pond, but you could also start with something simpler. The keys are moisture and hiding places. Plant a small area with native shrubs and grasses, add a pile of rocks or small boards on the ground among the plants to act as hiding spots, and keep the area wet with regular watering. Perhaps you already have an attractive situation in your yard that you don't even know about. A year ago, I noticed a very small leak in an irrigation line in my shed and thought about fixing it, until I looked down and noticed the crowd of Pacific Chorus Frogs and Western Toads staring imploringly at me from the moist haven I had unwittingly created for them. I did not repair the tiny leak, and my backyard wildlife are all the happier for it.

Protection: none

Warren Schmidt

AFRICAN CLAWED FROG

XENOPUS LAEVIS

Some alien movies are real doozies. *War of the Worlds* gave me nightmares for days, and my living room was decidedly not quiet during *A Quiet Place*. My mom tried to cover my four-year-old eyes during the chestburster scene in *Alien*, but I still watched it between her fingers and I don't think I have ever fully recovered.

But one of the most terrifying alien stories of all time is very real. Imagine a large aquatic monster, streamlined for swimming with torpedo-like ease, with a sense of smell underwater so keen that it can hunt in complete darkness. It seizes unsuspecting prey with its jaws and uses its sharp claws to tear the victim to pieces, then sucks its meal down in big chunks. This alien survives just fine if its watery home gets really cold or hot, is inundated with seawater, or even dries up. It can creep overland in the dark of the night to find new ponds to invade. It lays thousands of eggs that hatch into little babies with tentacles on their faces that they use to hunt until their sense of smell develops. The alien is itself impervious to disease but carries pathogens from pond to pond, where they infect and kill native wildlife.

This terrifying "alien" is the African Clawed Frog, a species native to Africa but established in numerous bodies of water throughout Southern California. How did they get here? Some populations were established from no-longer-wanted pets that were released into the wild by people before the species was banned in the California pet trade in the 1960s. But other populations have a more interesting history, involving pregnancy tests of all things! Scientists have known since at least the 1920s that pregnant women produce a hormone called hCG that is detectable in their urine. The old way of detecting this hormone (and thereby confirming pregnancy) was to inject a woman's urine into a female frog. If the urine had hCG in it, it caused the frog to lay eggs and the woman was declared pregnant. Doctors' offices had tanks of African Clawed Frogs that were basically living, swimming pregnancy tests. When someone later developed a blood test and "pee stick" to detect hCG, the frogs were no longer needed, and medical offices presumably released them into the nearest ponds. Ever since, this alien frog has been wreaking havoc on native species by eating them, outcompeting them for resources, and spreading disease.

While the idea of releasing a non-native species into the wild gives me heart palpitations, most people who do it think they are doing a kind thing for the animal. Many other species, like the Red-eared Slider (a turtle), have become established all over the country, even the world, when they outgrew their tanks and their owners set them free. While African Clawed Frogs and Red-eared Sliders will never be fully eradicated from California, their populations can be controlled by careful management.

An invasive African Clawed Frog in San Diego County. *Photograph by Jeff Lemm.*

If you see an African Clawed Frog in the wild, post photos and its location on iNaturalist so that scientists can follow up. The California Department of Fish and Wildlife also has an online invasive species reporting system.

Appearance: African Clawed Frogs have a very distinctive appearance, making them unlikely to be confused with any other species. They are large frogs with flattened torpedo-like bodies, round noses, eyes on top of the head, and smooth skin, all designed for ease of swimming. They have muscular legs and huge, webbed hind feet, and their unwebbed hands have long, thin clawed fingers. Their color ranges from green to tan, and they are usually blotchy. Eggs are laid singly or in small batches underwater. The tadpoles are very distinctive, with partially see-through bodies, long tails, and two small sensory tentacles on their faces that my students call "deely-boppers."

A group of recently metamorphosed African Clawed Frogs in San Diego County. *Photograph by Jeff Lemm.*

Natural History: African Clawed Frogs are almost 100 percent aquatic and are adapted for swimming, but adults can occasionally be found painstakingly loping overland during rainstorms, which is how these frogs spread to new wetlands. In California, these frogs mostly inhabit manmade or altered bodies of water, like irrigation ditches, cattle tanks, golf course water hazards, sewage lagoons, and the like. They are mostly nocturnal, so feeding and breeding typically happens at night. Males' calls are hard to describe, but they sound like a high-pitched recording of someone sawing a log. A male grasps a female around her waist with his head resting on her pelvis, then fertilizes her eggs as she lays them. Mating can happen anytime in California except for the dead of winter, with spring being the peak season. Females lay eggs one or a few at a time attached to underwater vegetation or rocks, and she can lay many thousands of eggs over an extended period of months. The eggs hatch rapidly and tadpoles metamorphose into small adults within a few months. African Clawed Frogs don't have tongues to grasp food like native Californian frogs do, so instead they bite onto prey with their jaws and force them down the throat, or if the meal is large then they use their sharp claws to dismember it first (I was not kidding about this alien being terrifying!). They eat basically anything they can find, alive or dead, including invertebrates, small fishes, and other amphibians, including young of their own species. The tadpoles filter feed on algae and other tiny organisms and dead stuff at the bottom of the pond.

Range and Variations: African Clawed Frogs are native to central and southern Africa. In California, they are established in Los Angeles, Orange, and San Diego Counties and have also been found in several surrounding counties, plus northern Baja California, Mexico, and Tucson, Arizona, and a growing list of other areas.

How to Find African Clawed Frogs: Walk around ponds (the more human impacted, the better) and look for large, big-footed frogs swimming or resting near the bottom. They are most active at night but can sometimes be seen at the bottom of the ponds during the day. You could theoretically jump in the water with a mask and snorkel to swim around and get a better look at African Clawed Frogs, but many of the ponds they inhabit have poor water quality (from fertilizers, cattle poop, human sewage . . . you get the idea), so I suggest that only for the most intrepid frog hunters.

Protection: none (non-native)

California Red-legged Frog. *Photograph by Spencer Riffle.*

RED-LEGGED FROGS

RANA AURORA AND *R. DRAYTONII*

In San Luis Obispo County where I live and work, there is only one native member of the "true frog" family, Ranidae. So, if we see a frog with the "typical" body shape of having a pointy nose, long legs, webbed feet, and smooth skin, we know it is a California Red-legged Frog. Well, we SLOcals *knew* its identity, that is, until American Bullfrogs spread into California sometime in the early 1900s, when they were brought from the eastern United States as a food source for fine restaurants, hungry gold miners in the Sierra Nevada, and other folks with a taste for sautéed frog legs.

Now, when we see that shape, we need to look more closely to distinguish the two species. There are a number of physical

California Red-legged Frog. *Photograph by Spencer Riffle.*

differences between native California Red-legged Frogs and American Bullfrogs, most obviously the presence of a large fold of skin along each side of the body in red-leggeds but not bull-frogs. But there is a fascinating behavioral dichotomy, too, that I will illustrate with an anecdote. On a field trip to a local area where both species can be found, I was helping one student free California Newts from nets we had dragged through the water when another student called from across the pond that he had found a frog. Which species, he wanted to know. Occupied with the newts, I called back to him, "Walk up to it and tell me what it does." Soon after, he reported that the frog let him walk right up to it and sat motionless, staring up at him. I called the class over to see this frog without even looking to confirm its identity, so sure I was that it was a rare California Red-legged Frog. These frogs have the unfortunate habit of allowing predators to approach closely, whereas invasive American Bullfrogs are very wary and dive to safety at the slightest threat. The presence of American Bullfrogs in California Red-legged Frog habitat is hugely problematic because they outcompete and eat the red-legged frogs and also can spread disease. The American Bullfrog is just one reason among many why the California Red-legged Frog is declining precipitously. The California Red-legged Frog's designation as California's official state amphibian does little to protect it from the impacts of the state's growing populace.

California Red-legged Frogs are an excellent, if unfortunate, illustration of how multiple stressors interact to negatively impact an amphibian species. In addition to the impacts of invasive species like American Bullfrogs, as well as plenty of fishes that eat eggs and tadpoles, California Red-legged Frogs have suffered from habitat destruction, overharvest as food, pollution, and disease. In the species account for California Red-legged Frogs on the excellent online field guide CaliforniaHerps.com, creator Gary Nafis writes with tongue-in-cheek flair that the frogs were first overhunted by "hungry alien predators known as the Miners of the California Gold Rush of 1849," only for most of the remainder to be killed off by "the insatiable apex predator, the Land Developer." Indeed, California Red-legged Frogs have been extirpated from most of their range, with small pockets of these frogs hanging on here and there. The Northern Red-legged Frog is in much better shape, but that is mostly because less of its habitat has been converted into shopping centers and housing developments.

The California Red-legged Frog also illustrates the importance of habitat protection in maintaining biodiversity. For them, it's really all about having clean freshwater habitats that are free from American Bullfrogs and non-native fishes. Unlike many other species of anurans, red-legged frogs appear to be rather resistant to the chytrid fungus. A biologist from a zoo that works on frog head-starting programs told me that their lab studies have shown that "it's almost impossible to infect a red-legged frog with chytrid." National Parks and zoo biologists recently restored wetlands, extirpated American Bullfrogs, and introduced head-started California Red-legged Frogs into the Merced River in Yosemite Valley where chytrid fungus had killed off other frog species. The introduced frogs thrived and have established self-sustaining populations. Success stories like this prove that it is possible to manage imperiled species, but it takes a lot of resources and human power to protect them.

Appearance: Both species of red-legged frogs are medium to large in size and often tan or gray in color with darker blotches along their backs. They have a prominent skinfold running from eye to hip on each side, a dark, banded pattern on the upper surface of their legs, and usually dark pink to bright red skin on the undersides of the legs. The best way to tell the two species apart is based on geographic location (see Range and Variations), but they share several subtle distinctions, too. The blotches on California Red-legged Frogs (*Rana draytonii*) usually have light centers, whereas the blotches are solid on Northern Red-legged Frogs (*Rana aurora*). Eggs are laid in batches of several hundred or thousand, all stuck together in an oval cluster attached to underwater vegetation, which sink in Northern Red-legged Frogs and remain closer to the water's surface in California Red-legged Frogs. Tadpoles are dark on top and light underneath, and covered in tiny, dark flecks.

Natural History: Red-legged frogs can be found near ponds, streams, and other water sources in forests and grasslands. They mate over a period of just a couple weeks at some time in the

Northern Red-legged Frog. *Photograph by Spencer Riffle.*

winter or early spring. During the daytime, males call for females from underwater or in the air, with Northern Red-legged Frogs tending to call underwater more often. Their calls are subtle chuckling sounds that we humans might not even be able to hear in many cases. Males grasp females by the armpits, then fertilize the eggs as the females lay them under the surface of the water. The eggs hatch in about a month, then tadpoles metamorphose within several months, although sometimes they can overwinter as tadpoles and metamorphose the following spring. Red-legged frogs eat almost any invertebrate or small vertebrate they can fit into their mouths, though insects dominate the diet.

Range and Variations: The Northern Red-legged Frog lives in Mendocino, Humboldt, and Del Norte Counties, then extends north through western Oregon, Washington, and southern British Columbia, Canada. The historic range of the California Red-legged

Frog extended from Mendocino County southward through coastal counties into Baja California, Mexico, plus the Sacramento Valley and the Sierra Nevada up to about 5,000 feet in elevation. However, most of those populations have now gone extinct, so the current range is spotty.

How to Find Red-legged Frogs: Red-legged frogs are diurnal (day active), so you can find them by simply walking slowly around the edges of ponds and other bodies of water in areas where they might occur, looking for frogs sitting in the water among aquatic vegetation or on the shore. You are most likely to find them in places where American Bullfrogs have not invaded. Listening for their mating calls is not a very effective way of finding them because they are quiet and sometimes call from underwater.

Protection:
Northern Red-legged Frog: species of special concern (California)
California Red-legged Frog: species of special concern (California), threatened (federal)

Harry Greene

RIO GRANDE LEOPARD FROG

RANA BERLANDIERI

If an organism with the name "Rio Grande" sounds out of place in California, that's because it is. The Rio Grande Leopard Frog is native to Texas, a bit of New Mexico, and a huge swath of Mexico south into Central America, yet this non-native species has been spreading through the southeastern part of the state for decades. Two practices are responsible for its introduction and subsequent spread in California: fish stocking and agricultural development. Since the Hoover Dam tamed the mighty Colorado River nearly one hundred years ago, the waters have been stocked with non-native fishes like Rainbow Trout, Channel Catfish, Largemouth Bass, and Bluegill to satisfy the appetite of sport fishermen. The Rio Grande Leopard Frog got into the lower Colorado River via fish

Jeff Lemm

stock that arrived from Texas or New Mexico, likely in the form of tadpoles that subsequently colonized the river. The frogs have slowly spread through the Imperial Valley ever since, facilitated by the water provided by agricultural fields. There once was a native Californian leopard frog that occupied the same area of California, the Lowland Leopard Frog (*Rana yavapaiensis*). This frog has not been seen in the state since the 1960s, making its extirpation coincident with introduction of the Rio Grande Leopard Frog as well as extensive habitat alteration. The irony of the situation is that the Rio Grande Leopard Frog in California looks almost exactly like the once native Lowland Leopard Frog—fascinatingly, these populations of invasive frogs have evolved to look more like Lowland Leopard Frogs than native-range Rio Grande Leopard Frogs in some ways since arriving in California, showing how the environment can shape evolution. So, while you might encounter leopard frogs if you go frogging among the lettuce farms of the Imperial Valley, these are definitively Rio Grande Leopard Frogs.

Appearance: Rio Grande Leopard Frogs are medium-sized frogs with pointy snouts and long legs used for jumping and swimming. As their name implies, they have leopard-like spots, usually dark green on a lighter green or tan background. They have folds of skin that run along each side of the body, plus a light stripe on each cheek. Their egg masses consist of several hundred eggs laid just below the surface of the water. The tadpoles have mottled dark and shiny gold colors.

Natural History: In California, Rio Grande Leopard Frogs are heavily associated with irrigation ponds and ditches in farmland. In their native range, they typically live and breed in streams. Rio Grande Leopard Frogs are active year-round except during cold winters, when they burrow into mud or hide under rocks alongside water bodies. Because these frogs are adapted to high temperatures, they can be found breeding almost continually throughout the year. The males call with a brief trill that they make a few times before stopping then repeating. Mating is typical froggy-style, with males grasping females from behind and fertilizing the eggs as the females lay them. Adults eat invertebrates and tadpoles graze on algae.

Range and Variations: Rio Grande Leopard Frogs are native to west Texas, southern New Mexico, and down through Mexico as far south as Nicaragua. In California, they occur in the Imperial Valley of Imperial and Riverside Counties and along the Colorado River in the extreme southeastern corner of the state.

How to Find Rio Grande Leopard Frogs: Rio Grande Leopard Frogs can be found active on warm nights and mild days by walking around the edges of agricultural fields where there is standing water in ditches or canals. If you want to encounter this non-native species, searching in the vicinity of El Centro and the Salton Sea are good bets. You can also listen for their calls by driving slowly through farmland at night. Take care not to trespass if you go searching for these frogs, as most farms do not allow unauthorized visitors and will probably assume people walking around at night with flashlights are up to something less wholesome than frogging.

Protection: none

FOOTHILL YELLOW-LEGGED FROG

RANA BOYLII

I am writing this account from my hotel room in Rohnert Park, where I am in town for a conference. There is a huge storm raging outside, so strong it has been dubbed an "atmospheric river." So, naturally I am heading out shortly with friends to look for sala- manders and frogs. I am right in the middle of "prime habitat" for Foothill Yellow-legged Frogs. From around this area northward into Oregon, it is still possible to see these frogs hanging out on the edges of their preferred rocky stream habitat. In much of the rest of their range in California, Foothill Yellow-legged Frogs have declined dramatically or gone extinct. Like many other frog species in California and elsewhere, this species has declined due to a large number of interconnected stressors. Much of their stream and river habitat has been destroyed or altered by water diversion and flooding, gold mining, logging, and cattle grazing. Fertilizers and pesticides from agricultural fields pollute their breeding waters, and chytrid fungus has caused die-offs. Non-native predators like American Bullfrogs and fishes eat the eggs and tadpoles.

We didn't see any Foothill Yellow-legged Frogs on our excursion during the big storm, but to be fair, most self-respecting frogs were hiding from the gale-force winds that day. We did see a small American Bullfrog on the road that tricked us into thinking he was a yellow-legged frog until we gathered around him and looked more closely. Why do invasive American Bullfrogs do just fine while native species suffer? That is a complicated question with many tendrils of answers, but one major one is that American Bullfrogs are not restricted to a narrow range of habitats. The plight of the Foothill Yellow-legged Frog illustrates what happens when multiple stressors come together to impact a species with very particular habitat requirements. The Foothill Yellow-legged Frog likes to breed and feed in specific types of water bodies—rocky streams and rivers. Only in the northwestern parts of our state, where rugged and dense forests have so far resisted deep penetration by human development, has this frog held on in decent numbers. This area of California harbors many sensitive species, not just the multitude of salamanders and frogs highlighted in this book but also salmon, owls, wolves, and mountain lions, not to mention insects, plants, fungi, and more. By protecting this area from logging and development, we can hold on to a large chunk of the diversity that the state of California boasts.

Appearance: Foothill Yellow-legged Frogs are medium in size with long legs and webbed feet that highlight their adaptations for jumping and swimming. Their skin comes in a range of colors, sometimes solid and sometimes mottled in color, and is covered in fine bumps like coarse grit sandpaper. The undersides of their thighs are covered in a light yellow, watercolor-like wash. They usually have a pale stripe across the tops of their heads, extending from one eye to the next. The egg masses are laid in clusters of about a thousand eggs and can sometimes be hard to see as they become covered in silt. The relatively large tadpoles are brownish and speckled to blend in underwater and have eyes on the tops of their heads like flounders.

Natural History: Foothill Yellow-legged Frogs breed in rocky rivers and streams (and, very rarely, in ponds or mud-bottomed creeks), and are typically found not far from these waters in forested area or chaparral, especially if there are sunny clearings.

They are active mainly during daytime hours. Males don't call much because they mate in rivers where the loud, flowing water would mask the sounds, and when they do call it is mostly underwater. Occasionally you might hear one calling on land. The sound is hard to describe . . . it is rather like a short, squeaky grunt. Mating typically occurs in spring, with smaller males grasping larger females by the armpits from behind. The females lay batches of eggs attached to the downstream sides of rocks in shallow water as the males fertilize them. A fascinating observation described one male forcefully kicking another male off a female just as she laid her eggs; the new male proceeded to swoop in and fertilize them. The naturalist who observed this became my hero when they dubbed this behavior "clutch piracy." Eggs typically hatch within a few weeks, then the tadpoles feed on aquatic algae and decomposing matter and metamorphose several months later, in late summer or fall. Adult Foothill Yellow-legged Frogs feed on a wide array of invertebrates and occasionally cannibalize young frogs.

Range and Variations: The Foothill Yellow-legged Frog's historic range included much of coastal California from Ventura County north into central coastal Oregon, plus the lower reaches of the Sierra Nevada (hence the name "Foothill") and disjunct populations here and there, including Los Angeles County and one in the mountains of northern Baja California. However, populations of Foothill Yellow-legged Frogs south of Monterey County have apparently mostly gone extinct, and populations in the Sierra Nevada are declining rapidly. Foothill Yellow-legged Frogs are classified overall as a California species of special concern, so they are protected from being captured or handled.

How to Find Foothill Yellow-legged Frogs: The best way to find Foothill Yellow-legged Frogs is to walk slowly along the banks of open, sunny, rocky rivers and creeks during the day in coastal Northern California, watching for frogs basking along the edges of the water. When they detect your presence, they will jump into the water, swim to the bottom, and hide among the rocks. If you search for Foothill Yellow-legged Frogs in late summer, you can sometimes find hundreds of newly metamorphosed froglets hanging around in the small puddles left behind when the streams begin drying up. Listening for frogs calling at night is not very fruitful because males don't chorus for females in the same way that many other frog species in California do.

Protection: status depends on location; most are threatened or endangered at both California and federal levels

Spencer Riffle

CASCADES FROG

RANA CASCADAE

If you are reading the species accounts in this book in order, by now you might be starting to grow a bit depressed about the status of California's amphibians. So many frogs are in trouble in particular due to habitat destruction and disease, among other problems. While the Cascades Frog is no different in this respect, I am going to tell you a fascinating story about research into their declines. Stories like this help me to maintain a positive outlook about the plight of California's amphibians because they highlight the ingenuity and passion of the scientists working to save them.

Cascades Frogs, like many other species, used to be extremely common around high-elevation ponds and streams in

Northern California. Biologists in the 1920s wrote in their field notes of "a frog every yard" as they walked around mountain lakes. That is no longer the case. The Cascades Frog is extinct in many areas, and numbers of frogs are much lower in the areas they still live, especially in the southernmost populations of their range, which happen to be those in California. Why?

Scientists have used a fascinating method to suss out one of the major causes of the decline of California Cascades Frogs: looking for pathogen infection . . . in dead, pickled frogs from decades ago. Natural history museums around the world house millions of preserved specimens of animals, plants, fungi, and other forms of life. Scientists of the past collected these specimens and deposited them in museums with no clue as to what scientists of the future would use them for; they only knew that they might need them for something important. They could not have dreamed how important the Cascades Frogs that they painstakingly preserved and placed into glass jars would end up being.

About a decade ago, scientists analyzed skin samples from Cascades Frog museum specimens that had been collected starting in 1907. The scientists used a technique called the polymerase chain reaction (PCR) to detect the presence of the genetic material of the chytrid fungus in the frogs' skin. This same technique is used to detect chytrid in living frogs today by swabbing their skin, but I find it amazing that we are able to use it on one-hundred-plus-year-old pickled frogs! They found that specimens collected prior to the 1970s had no chytrid but that sometime in the mid-1970s infected frogs started showing up in collections. This coincided with the beginning of reported Cascades Frog declines.

The chytrid fungus is not the only problem the Cascades Frog faces—they also suffer from degradation of their high-elevation meadow habitat, exposure to increased levels of UV radiation, introduced fish predators, and other stressors. However, the

Spencer Riffle

chytrid fungus study was incredibly important because it confirmed that major declines coincided with the arrival of a deadly pathogen. Unlike fish, which can be removed, and habitat, which can be restored, the chytrid fungus is in the water to stay. Reintroducing Cascades Frogs will require more research and creativity, like some of the methods currently being used to successfully repatriate yellow-legged frogs (see page 198).

One more thing: this study highlights the importance of museum specimens. Many people do not like the idea of collecting animals to euthanize them and place them into jars for a private natural history museum. However, without continued collection of such specimens, there will not be a record frozen in time for future scientists to use to help solve problems that we cannot yet anticipate will arise.

Appearance: Cascades Frogs are medium sized, with smooth skin, long legs, prominent folds along the sides of the body, and somewhat webbed hind feet. They are typically brown or greenish with small, dark spots on their backs. A dark, triangular patch extends from the outer edge of each eye to the upper arm. The egg masses are baseball sized and consist of many small, dark eggs separated by thick jelly coating, and they are laid in shallow water. Tadpoles are dark brown with speckling.

Natural History: Cascades Frogs are found near bodies of water in high-elevation meadows and open areas in pine forests, where they sit on the water edges during the day and feed on insects. In late spring and summer, males sit in the water and call for females both day and night with a sound like a chicken clucking, then grasp them from behind and fertilize the eggs as the females lay them. The embryos develop rapidly and hatch into tadpoles that also metamorphose rather rapidly, growing legs and leaving the water later that summer.

Range and Variations: Cascades Frogs currently can be found in high-elevation areas in Plumas, Shasta, Trinity, and Siskiyou Counties, and tiny slivers of adjacent counties. They also range through parts of Oregon and Washington up to extreme southern British Columbia, Canada.

How to Find Cascades Frogs: Walk around mountain lakes, ponds, and streams during the day from late spring through early fall, looking for Cascades Frogs sitting along the edges. They will jump into the water with a loud plop and swim across the pond or to the bottom of the water. You are not permitted to capture or otherwise interfere with them, so be sure only to admire them from a distance.

Protection: species of special concern (California)

Max Roberts

AMERICAN BULLFROG

RANA CATESBEIANA

~~~~~~

FAMILY RANIDAE

Ah, the mighty American Bullfrog. What a beast! A veritable hopping stomach. Able to disperse overland as far as one kilometer in one night. Perhaps the most hated amphibian of the West. Why? Through no fault of its own, other than having the audacity to sport large and delicious legs that people love to eat, the American Bullfrog was brought west by gold miners, chefs, and other folks, then released to establish breeding farms. These frogs proceeded to wreak absolute havoc on wildlife in just about every corner of

California, apart from high-elevation spots and some dry areas of the deserts where the frogs have failed to establish populations. They eat anything and everything, and they can carry and spread diseases to native wildlife.

Although there are too many American Bullfrogs in California to eliminate them entirely, people spend a lot of time eradicating them from sensitive areas where threatened and endangered species persist. My students and I spent several years eradicating bullfrogs from a private property in Santa Margarita by painstakingly dragging nets through ponds and sorting through tadpoles to separate the protected native California Red-legged Frogs from the invasive American Bullfrogs. For about ten years the property was bullfrog-free, until 2024 when the owner called me to report that he heard the dreaded "vrooooom vrooooom" sound of a male bullfrog calling from one of the ponds. My students were on the case and worked tirelessly to capture the trespasser and help protect the resident California Red-legged Frogs. This incident highlights how difficult it is to manage these invaders of the American West, as you must stay on top of it indefinitely to keep future invaders away. What should you do if you encounter American Bullfrogs in California? There is no good answer, as captured bullfrogs should be euthanized, but it is important to do this in an ethical way. The California Department of Fish and Wildlife recommends reporting bullfrog encounters; search online for their invasive species reporting form, email, and phone hotline, and choose one of the ways to alert them to a population of American Bullfrogs.

Appearance: American Bullfrogs are large, muscular frogs with long legs and huge, webbed hind feet suited to their jumping and swimming lifestyle. They are usually green or brownish. They have a very obvious round circle behind each eye called the tympanum

Max Roberts

that helps transmit sounds to the inner ear underneath. The tympanum is larger in males than in females. Females lay huge masses of thousands of eggs that float on the surface of the water then sink after a few days, just before hatching. The tadpoles are large, greenish, and spotted, and can often take one or more years to metamorphose.

Natural History: American Bullfrogs inhabit just about any body of fresh water they can find. Irrigation canals, reservoirs, creeks, backyard pools—all of these are good habitats. During the cold of winter, they bury themselves in the mud underwater and then emerge to mate and feed in the spring and summer. Males often call for females from the edges of the water, where they can jump into the depths to escape any predators (or herpetology students) that approach. Their call sounds like a deep "vrooooom." After pairing up for a while with the male wrapping his arms around the female's armpits from behind, she lays a huge raft of eggs that he fertilizes. American Bullfrogs can be active at all times of year except cold winters. They eat literally anything that moves, including invertebrates and vertebrates like mice, birds, and other frogs, even other bullfrogs. Tadpoles eat mostly algae and plants.

Range and Variations: The American Bullfrog is native to the eastern United States. The species has spread throughout much of world from people releasing them to establish populations for food. In California, American Bullfrogs can be found anywhere in the state except for high elevations in the mountains and in parts of the southeastern deserts, though they thrive at desert oases and will readily establish populations if they have a means of getting there.

How to Find American Bullfrogs: American Bullfrogs are large and therefore conspicuous. They can be active at any time of day, including warm summer nights. Walk along the sides of ponds and other bodies of water, watching for large frogs sitting on or near the edges of the water. They will often jump into the water to get away from you, sometimes squeaking and making a large, plopping noise as they bellyflop. You can hear them calling from aquatic areas throughout much of the year, making it easy to detect them even if thick aquatic vegetation makes it difficult to see them. Methods for capturing American Bullfrogs are diverse and beyond the scope of this book, but common techniques include using nets and hand-catching them, and recently scientists have developed traps that target bullfrogs.

Protection: none (non-native)

Sierra Nevada Yellow-legged Frog. *Photograph by Spencer Riffle.*

MOUNTAIN YELLOW-LEGGED FROGS

RANA MUSCOSA AND *R. SIERRAE*

FAMILY RANIDAE

When I took herpetology at UC Berkeley back in 1997, the Southern Mountain Yellow-legged Frog and the Sierra Nevada Yellow-legged Frog were still considered a single species and not yet endangered. But the events leading to these frogs' peril were already well underway. Once extremely common throughout much of the Sierra Nevada and Southern California mountain ranges, populations of yellow-legged frogs were declining due to the combined effects of climate change, invasive species, disease, pollution, and more. The populations further north and at higher elevations had an additional problem, too—tadpoles can take three to four years to metamorphose to adults, so as global warming reduced the snowpack in the Sierra Nevada and therefore the subsequent snowmelt, a single event of a pond drying up meant that three to four generations of frogs blinked out.

As if these circumstances weren't bad enough, eggs and tadpoles were being gobbled up by fishes. This was happening way up in remote high-elevation lakes in the Sierra Nevada, where no native fish had ever swum but where non-native fishes were stocked by the National Park Service, California Department of Fish and Wildlife, and other groups to create recreational fishing opportunities. In the 1990s, a graduate student at UC Berkeley named Vance Vredenburg (who, coincidentally, took herpetology with me or was the teaching assistant—I can't recall) conducted a study with a grand goal: to manually remove all introduced fishes from a series of remote Sierra Nevada lakes to observe whether the yellow-legged frogs recovered. Vredenburg and a team of undergraduates backpacked miles into the wilderness at the start of the summer (there are no roads leading to these lakes), set up camp, and got to work. Despite the naysayers who doubted they could catch all the fishes, catch them they did—I like to think about all the delicious grilled meals they enjoyed that summer. The experiment was a resounding success. Over the course of a few years of fish removal, the frog populations bounced right back. As a direct result of this research, fish are no longer stocked in many parts of the Sierra Nevada.

Disturbingly, some populations of yellow-legged frogs have gone extinct in recent decades, even though no fishes were present in their lakes. The culprit was identified as the chytrid fungus,

Sierra Nevada Yellow-legged Frog. *Photograph by Spencer Riffle.*

the same pathogen that has decimated amphibians worldwide. In Yosemite National Park, many but not all populations of Sierra Nevada Yellow-legged Frogs that had already been stressed by climate change and invasive fishes ended up dying off due to the fungus. Could the populations that did not die off hold clues to how we could restore frogs throughout the landscape? Researchers recently reported the results of a fifteen-year study reintroducing frogs from the seemingly resistant populations to areas where they had died off throughout the park (after fishes were removed, naturally). Though the fungus is still present, the imported frogs are thriving. It's not just about the frogs, of course. It's about the recovery of an entire ecosystem in Yosemite. About the successful reintroductions, lead researcher Roland Knapp said, "You sit on the bank and you have tadpoles all around you in the water and adult frogs sitting next to you on the shore. You have birds flying in and feeding on them, and snakes that are feeding on them. You have a lake that's alive again."

The outlook is bleaker for Southern Mountain Yellow-legged Frogs. Most of their populations have blinked out, and the heroic breeding, rearing, and research program led by the San Diego Zoo Wildlife Alliance (see page 20) may be the species' sole lifeline. So far, scientists have not found any naturally chytrid-resistant frogs that could be used to repatriate their habitat, like they have with Sierra Nevada Yellow-legged Frogs. Could one species of yellow-legged frogs hold the key to saving the other species? What makes some Sierra Nevada Yellow-legged Frogs resistant to the devastating fungus but no Southern Mountain Yellow-legged Frogs so lucky? The answer is likely complex, and I hope that it will be answered by scientists of the future. In the meantime, we owe our thanks to zoos for acting as a Noah's Ark of sorts for California's endangered amphibians.

Appearance: These are medium-sized frogs with mottled brown or green markings on a yellow or cream background. They have long, yellow legs with webbed hind feet for swimming. Though the Southern Mountain Yellow-legged Frog (*Rana muscosa*) has somewhat shorter legs than the Sierra Nevada Yellow-legged Frog (*Rana sierrae*), the best way to determine which is which is by location (see Range and Variations).

Natural History: These frogs are highly aquatic, found in or near mountain lakes, ponds, and streams. They overwinter underneath the ice and mate when the ice and snow melt in the spring (southern and lower-elevation populations) or summer (northern and higher-elevation populations). The males call for females underwater, then grasp them from behind and fertilize the eggs as the females lay them in clusters in shallow water. Southern California populations of Southern Mountain Yellow-legged Frogs may hatch and metamorphose later that summer, but tadpoles from their northern populations as well as most Sierra Nevada Yellow-legged

Southern Mountain Yellow-legged Frog. *Photograph by Max Roberts.*

Frogs don't usually metamorphose for three or four years. The diet consists of invertebrates and small tadpoles.

Range and Variations: The Sierra Nevada Yellow-legged Frog historically occurred throughout the Sierra Nevada from Plumas County south to Inyo County, though they now have a spotty distribution in that area. The Southern Mountain Yellow-legged Frog also has spotty populations in the southern Sierra Nevada (mostly Fresno and Tulare Counties), plus small, imperiled populations in mountains of Riverside, San Bernardino, and Los Angeles Counties.

How to Find Mountain Yellow-legged Frogs: The best way to find these frogs is to walk around mountain lakes, ponds, or streams during the day in the summer, watching for frogs sitting near the edges of the water. If you get too close, they will escape into the water.

Protection:
Southern Mountain Yellow-legged Frog: endangered (California, federal)
Sierra Nevada Yellow-legged Frog: threatened (California), endangered (federal)

Northern Leopard Frog. *Photograph by Jeff Lemm.*

NORTHERN AND SOUTHERN LEOPARD FROGS

RANA PIPIENS **AND** *R. SPHENOCEPHALA*

I was an odd teenager. You might have had an opportunity, like I did, to dissect a preserved frog in your high school science class. That frog was likely a leopard frog, almost certainly a Northern Leopard Frog, collected somewhere near the center of our country, sold to a biological supply company, then sent on to your local high school for the purposes of educating our youth. You probably threw your frog into the waste bucket when you were done examining its organs, as a normal human being would do. Not me. In Mrs. Derr's tenth-grade biology class at Novato High School, after picking through my frog's organs and eggs with care (I was lucky and got a female frog), I surreptitiously stashed the frog's abdomen in my desk. I wanted to keep looking at the organs for longer than just one class period. Unfortunately, the abdomen shriveled up over the course of a few days and made my desk start to stink, so I ended up tossing it before Mrs. Derr found it. A good thing, really, as hoarding a dead frog might have earned me a referral to the school counselor. But my curiosity about leopard frogs now extends beyond their intestines and into the bizarre story of the current state of California's leopard frogs.

Chances are that if you encounter a leopard frog here, it will not be native . . . even if it's technically a native species. Let me explain. Northern Leopard Frogs historically occupied several areas in eastern California, representing the westernmost extent of their vast range. Over the past century, many of these populations have gone extinct, probably due to a combination of disease and exotic predators like American Bullfrogs, fishes, and crayfish. The complicating factor is that all the while, surplus biology class leopard frogs were being released throughout the state. Adding to the confusion is the fact that some of the released frogs might have been Southern Leopard Frogs, which are native to the southeastern United States and used to be considered the same species as Northern Leopard Frogs. So, California now houses several populations of leopard frogs of unknown origin, with few known populations of native Northern Leopard Frogs. For this reason, Northern Leopard Frogs are protected in California, even though most of them are not native.

Appearance: Leopard frogs are medium-sized frogs with long, muscular legs, webbed feet, pointy snouts, and large, dark spots on a tan, green, or gray background. They have defined ridges of skin running down each side of the body from the eye to the hip. Though it can sometimes be difficult to tell the two species apart, in general Northern Leopard Frogs (*Rana pipiens*) have more dark spots stippling their sides, the dark spots are more starkly outlined in white, and skin ridges are wider and more distinct than in the non-native Southern Leopard Frogs (*Rana sphenocephala*). Eggs are laid in large, pancake-like clumps underwater and hatch into brownish tadpoles with shimmery gold spots on their backs. The young froglets don't have many spots and develop these as they grow.

Natural History: Most natural history information is based on populations occurring outside California. Leopard frogs can be found in many types of aquatic habitats, ranging from rivers to ponds to irrigation canals, and other artificial bodies of water. Their prime habitats are large, shallow, permanent ponds with plenty of aquatic vegetation to hide among. Mating typically occurs in the spring and summer. Males sit in the water and call for females at night. In both species, the calls are long, irregular, guttural, growly sounds. Males grasp females from behind to fertilize the eggs as the females lay them in large, flattened piles under the surface of the water, usually attached to vegetation. The eggs hatch rapidly, and the tadpoles feed on aquatic algae and decomposing organisms, and usually metamorphose into froglets within a few months. Adults eat insects and other invertebrates, plus the occasional small vertebrate.

Range and Variations: The native range of Northern Leopard Frogs in California consists of a spotty distribution in the northern and

A Southern Leopard Frog in Georgia. *Photograph by Noah Fields.*

eastern parts of the state. From there, they range very widely throughout the Plains states, northeastern United States, and Canada. In California, leopard frogs are established in Fresno, Mono, Merced, Madera, and Riverside Counties, and there are likely other populations as well.

How to Find Northern and Southern Leopard Frogs: As described earlier, Southern Leopard Frogs are non-native to California, and most of the extant populations of Northern Leopard Frogs are also introduced. However, because the Northern Leopard Frog is a rare native species, it is protected. If you decide to search for either of these species, you'll need to do some research on exactly where they are currently documented to occur, since it is not my goal in this book to give specific locations to those seeking to find frogs. Walk around edges of ponds, irrigation ditches, and slow-moving streams in areas where these frogs

A Northern Leopard Frog in Minnesota takes shelter in the shade of a rock.
Photograph by Owen Bachhuber.

are known to occur. During the day, your presence might scare a frog into jumping from its concealed position among the vegetation on the bank of the water; they sometimes emit a loud squeak as they plop. You can also search for these frogs on warm spring or summer nights by listening for the males' calls and shining your flashlight into the water.

Protection:
Northern Leopard Frog: species of special concern (California)
Southern Leopard Frog: none

Spencer Riffle

COUCH'S SPADEFOOT

SCAPHIOPUS COUCHII

~~~~~~

FAMILY SCAPHIOPODIDAE

The adorable spadefoot is a good candidate for melting the hearts of any holdouts who are still resistant to the charms of Californian amphibians. These frogs have enormous eyes and snub noses, and large mouths that seem permanently frozen in a slightly grumpy expression. The tiny limbs of these frogs tell you that they aren't good jumpers or swimmers. Instead, spadefoots use the namesake "spades" on their hind feet to dig themselves underground, where they secrete a water-resistant cocoon around their bodies and sit there for months, sometimes years, until the next rains arrive. Like all spadefoots, the Couch's Spadefoot is primarily a *fossorial* species, meaning that it spends

A female (left) and a male (right) Couch's Spadefoot. *Photographs by Max Roberts.*

almost the entire year underground and emerges only for a few weeks to breed and feed before digging itself back underground. If this isn't amazing enough, it's the tadpoles that are the biological stars of the show. Because Couch's Spadefoots mate and lay their eggs in temporary pools of rainwater, their tadpoles must develop quickly to avoid dying if the water evaporates. This species holds the record for the fastest development of all frogs in North America—it can go from an embryo in an egg to a tadpole to a tiny froglet with four legs crawling onto land in just one week! These adorable amphibians are protected in California partly because they barely range into the southeastern part of the state and partly because that area of California has experienced fair amounts of agricultural development that has altered spadefoot habitat. However, where they do occur, they can be extremely abundant. I teach a field herpetology course in Arizona and New Mexico in the summer, and the monsoon storms bring thousands of these little frogs to the surface, resulting in a veritable spadefoot obstacle course when night driving on remote roads.

Appearance: Spadefoots are sometimes called "spadefoot toads" because they are somewhat reminiscent of true toads in the family Bufonidae. However, they are only distantly related to toads. Spadefoots are small anurans with short arms and legs, large bulbous eyes, and warts on their skin, but no glands behind their eyes. They have a conspicuous dark spade on each hind foot. The sexes usually differ in color and pattern, with males being light green with faint dark blotching, and females having a strong mottled brown and light green pattern. Eggs are laid in clumpy strings on underwater vegetation. The tadpoles have a golden sheen and rapidly metamorphose into small froglets.

Natural History: Couch's Spadefoots are the most arid adapted of all of California's amphibians. They occur in deserts and grasslands, spend most of the year underground, and emerge during heavy spring and summer rains to mate and feed. Males call for females from ephemeral ponds made of rain or flood water. The sound of a male Couch's Spadefoot call is hard to describe. To me, it sounds like a highly nasal cat's meow. When a female selects a male, he grasps her around her waist from behind, and she lays strings of thousands of eggs as he fertilizes them. The eggs hatch within a single day, then the tadpoles graze on algae and are capable of rapidly metamorphosing, especially if the pond is very small. If the water dries up before the minimum one week necessary for complete development, all the tadpoles will desiccate and perish in a pile called spadefoot "popcorn" or "brittle." Couch's Spadefoot adults feed on arthropods and other invertebrates in the short window of time that they are active. This provides them with enough stored fat to survive on while hiding quiescent underground for the next year.

Range and Variations: Couch's Spadefoots range through the southwestern United States from extreme southeastern California to Texas and also into northern Mexico. In California, they are mainly found in eastern Imperial and Riverside Counties.

How to Find Couch's Spadefoots: A good way to find Couch's Spadefoots is to employ a trick you may know about if you have learned how to search for nocturnal snakes in the desert: night driving. Drive slowly (25–40 mph, depending on how good you are at seeing small animals on the road) through good habitat within the Couch's Spadefoot's range on warm spring or summer nights during or right after a storm. With luck, you will see lots of these little frogs sitting around on the asphalt at night. You can also drive with your windows down and listen for the distinctive meow-like call of male Couch's Spadefoots, then hike to find the ponds where they are breeding.

Protection: species of special concern (California)

Spencer Riffle

WESTERN SPADEFOOT

SPEA HAMMONDII

FAMILY SCAPHIOPODIDAE

In 2007, I taught herpetology at Cal Poly for the first time. My students and I visited a private ranch in the foothills of the Santa Lucia range in the Carrizo Plain, where we spent the day catching lizards and hunting for rattlesnakes. At night, when it got too cold for reptiles to be out and about, we donned our rubber boots and headed out to look for frogs. It had recently rained, so the Western Toads were out in force and tiny male Pacific Chorus Frogs sang their hearts out from the thickly vegetated creeks until our troupe of noisy humans arrived. I asked the students to gather around, turn off their flashlights, and stand silently. As soon as the lights went out and the conversation died down, the frogs cautiously resumed their amorous chorus. We stood in complete darkness, enjoying the experience of a sound-only sensory world. That was when I heard a loud, on-and-off, snore-like buzzing start up from a nearby cattle pond. That sound meant only one thing: Western Spadefoots! We rushed over to the pond and saw a handful of male spadefoots speckling the pond, yelling hopefully for females. They sucked in a breath, then their chins billowed out as they emitted a loud, snore-like buzz, and they repeated this over and over. I gushed at the students about how lucky they were to see this species in the wild, as the Western Spadefoot is the most difficult frog to find in the area of California where we live, and that is saying something because its scarcity even one-ups the highly endangered California Red-legged Frog. Why? As a fossorial species, Western Spadefoots spend most of the year underground. Then, boom! They burst from the soil, mate, and feverishly gorge themselves on bugs before digging themselves back underground, where they stay until the following year's rains. While the sound of the rain hitting the ground is thought to be the stimulus to dig themselves out, no one knows for sure why they emerge during some storms but not others. Returning to a spot where you previously found them is a good strategy, but it doesn't always work. We regularly search that cattle pond and surrounding areas every spring, sometimes right after a rainstorm, and in nearly twenty years I have only seen spadefoots there one more time. These adorable animals are protected in California, where much of their habitat has been converted into farmland. In 2023, the species was proposed to be listed as threatened under the US Endangered Species Act. I hope that such measures will help preserve this incredible animal so that future generations of amphibian hunters will be able to experience the buzzy snore of the diminutive yet mighty Western Spadefoot.

Marisa Ishimatsu

Appearance: Western Spadefoots are small to medium in size, with stout bodies and large, bulbous eyes with vertical pupils that widen into ellipses in the dark. They have a mottled pattern, often of greens, browns, and light grays. They have very short legs, reflecting their lifestyle of digging rather than jumping. The conspicuous black spade on each of their hind feet gives them their name. Eggs are laid in small clumps attached to vegetation under the surface of the water. After hatching, the larvae start out small and grow rapidly into stout, iridescent brownish tadpoles before metamorphosing.

Natural History: Western Spadefoots can be found in a variety of habitats including grasslands, chaparral, and forests, though they are most commonly found in areas with little or no tree cover. Like all spadefoots, Western Spadefoots spend the majority of the year hiding deep underground in a sort of stasis. Strong late

Spadefoots use spades on their hind limbs to dig themselves underground.
Photographs by Spencer Riffle.

winter and spring rains stimulate them to crawl to the surface, where they breed in temporary pools of water. Males attract females with buzzing calls that sound like a cross between a snore and a sped-up woodpecker's peck. A male grasps a female around her waist, then deposits sperm on eggs as she lays them. Embryos develop and hatch in less than a week, and tadpoles feed on algae, decomposing matter, and sometimes other Western Spadefoot tadpoles, then metamorphose in one to three months. Adults feed on arthropods during this brief period aboveground, then use the spades on their hind legs to dig back down into the ground. In areas with little rainfall, spadefoots may not emerge at all during a given year. Because it's difficult to study animals that live underground, scientists know little about the survival of spadefoots that fail to emerge.

Range and Variations: Western Spadefoots live in coastal counties in southern and central California up to Monterey Bay, plus they historically occupied much of the Central Valley well into Northern California, where they are now found only in disjunct populations. They also extend from Southern California down the Pacific Coast into northern Baja California.

How to Find Western Spadefoots: Most of the year, it's not possible to find spadefoots because they are buried deep underground. The key to finding them is to go out looking for them within a few days of a major storm in the late winter or spring. I drive along roads at night in areas with rain-filled roadside ditches, listening for the distinctive calls of males trying to attract females. In California, you are not permitted to touch these frogs, which may be just as well because when harassed they secrete a substance onto their skin that can cause severe hay fever in people who make the mistake of touching their faces after handling a spadefoot. This is too bad because it can otherwise be fun to take a sniff of Western Spadefoots: they smell like garlic peanut sauce!

Protection: species of special concern (California)

Ryan Sikola

GREAT BASIN SPADEFOOT

SPEA INTERMONTANA

FAMILY SCAPHIOPODIDAE

We humans are adapted to connect scents with wonderful memories from our lives. Many people have fond memories associated with the smell of cookies baking in Grandma's kitchen, or the balsam scent of Christmas trees, or the thick coconut aroma of sunblock on a beach day. My favorite scent is associated with wonderful memories that are tied up with horizon-spanning rainbows, the rushing of water through dry washes, and the promise of frogs. This smell is desert rain. The molecules that are aerosolized by water impregnating sun-hardened soil and the dry leaves of sage and creosote bushes create the most delicious scent nature has in her coffers. Perfume and incense makers have tried to capture the essence of desert rain, but so far I have not found anything that comes close to the real thing.

Perhaps the best part of the smell of desert rain is what it signifies: an entire community of desert animals simultaneously arousing from their dehydration-induced stupors and coming out to drink . . . and play. Desert amphibians like the Great Basin Spadefoot spend all year patiently waiting for rain before finally emerging from the soil to breed and feed joyously in the warm, flooded desert plains. The water also brings out their prey (ants and beetles) and their predators (snakes, coyotes, owls, and more), creating a veritable smorgasbord in the desert that trumps the most decadent Las Vegas buffet. If you have never had the pleasure of witnessing a rainstorm in our state's arid lands, make that your resolution this year. Pack up your field guides and your copies of *California Snakes and How to Find Them*, *California Lizards and How to Find Them*, and *California Amphibians and How to Find Them*—and might I recommend adding Obi Kaufmann's *The Deserts of California*?—and head east. Get ready for the adventure of a lifetime!

Appearance: Great Basin Spadefoots are small to medium-sized frogs with features typical of this family, including wedge-shaped black hardened spades on their hind feet, warty skin, and beautiful protruding golden eyes with no glands behind them. They have a blotchy gray-and-brown pattern, sometimes with tiny orange warts, and distinctive spots of dark eyeshadow on each eyelid. The jelly-like eggs are laid in small clusters attached to underwater vegetation. Tadpoles are dark on top and white underneath.

Natural History: Great Basin Spadefoots are adapted to deserts. They spend most of the year underground, cocooned up to prevent water loss to the dry soil. Spring and summer rains bring them to the surface for a flurry of mating and gorging on insects, especially ants, that are also stimulated by the rains. Like other members of their family, Great Basin Spadefoots specialize in breeding in temporary pools of rainwater, but in the eastern deserts of California where rain can be rare, they also mate in desert springs and irrigation ditches. Males holler at night from the water to attract females with loud, buzzing, quack-like calls, then wrap themselves around the waists of approaching females to position themselves to fertilize the eggs. Females lay the eggs in small clumps in shallow water. The eggs hatch within a few days into small, dark tadpoles that scavenge dead things in the water and feed on aquatic vegetation or small animals, sometimes including other Great Basin Spadefoot larvae. The tadpoles metamorphose in one to two months.

Range and Variations: In California, the Great Basin Spadefoot is found only in desert regions in the northeastern corner of the state and in the northern Owens Valley mainly in Mono and Inyo Counties. From there, it extends northward through the deserts of eastern Oregon and Washington into southern Canada, east throughout Nevada and Utah into western Wyoming and Colorado, and south into northern Arizona.

How to Find Great Basin Spadefoots: If you have read the accounts for the Couch's and Western Spadefoots in this book, you probably have an idea about how to find the Great Basin Spadefoot: drive through the desert in its range at night following spring or summer rains with your windows down, listening for their drawling, duck-like quacks, then follow the sounds on foot with a flashlight to find their mating ponds. Great Basin

Ryan Sikola

Spadefoots can sometimes be found in desert springs, irrigation ditches, and other bodies of water in the desert, even when it hasn't recently rained. Handle Great Basin Spadefoots with care; their skin releases a chemical that smells pleasantly of popcorn but loses its pleasantness when accidentally rubbed into one's eyes or nose, where it has been known to cause long strings of alternating sneezing and swearing.

Protection: none

ACKNOWLEDGMENTS

This book involved lots of "research." By that, I mean excuses to go herping for amphibians in the winter. Big thanks to my graduate students and herping friends for the company. I thank Robert Hansen and Brandon Kong for reviewing a draft of this book. The websites AmphibiaWeb.org and CaliforniaHerps.com were important references for this book, as was Robert Hansen and Jackson Shedd's new field guide *California Amphibians and Reptiles.* Stephanie Chancellor from the San Diego Zoo Wildlife Alliance and Rochelle Stiles from the San Francisco Zoo provided valuable insight into zoo programs. My editors, Emmerich Anklam and Marthine Satris, and the rest of the Heyday team helped shape the book into its final form. The best part of this book is its photographs, which the following photographers graciously allowed me to use: Owen Bachhuber, Sean Barefield, Ralph Cutter, Marisa Ishimatsu, Dante Fenolio, Noah Fields, Zeev Nitzan Ginsburg, Rob Grasso, Harry Greene, Francesca Heras, Doug Hofmockel, Ceal Klingler, Brandon Kong, Jeff Lemm, Matthias Lemm, Jeff Martineau, Spencer Riffle, Max Roberts, Mike Rochford, Warren Schmidt, Jackson Shedd, Ryan Sikola, Dustin Smith, and Alexander Yan. My parents, brother and sister, and husband, Steve, are the best cheerleaders a girl could ask for, and I am so grateful for their constant encouragement. Finally, I want to double-down on my dedication of this book to my students and mentees, who in the process of learning with me have taught me more than they could imagine.

RECOMMENDED FURTHER READING

Everybody loves amphibians, and hopefully after reading this book you love them even more! Here is a list of resources for learning more about the frogs, toads, and salamanders of California and beyond.

FIELD GUIDES

Hansen, Robert W. and Jackson D. Shedd. *California Amphibians and Reptiles* (Princeton Field Guides). Princeton University Press, 2025.

California finally has a dedicated field guide! This one is thorough, accurate, and an absolute must-have for the serious field herpetologist.

AmphibiaWeb.org

AmphibiaWeb is a comprehensive resource for detailed scientific information and news about the thousands of species of amphibians around the world.

CaliforniaHerps.com

I highly recommend this excellent free, online field guide to the amphibians and reptiles of California, created by Gary Nafis.

BOOKS

Duellman, William E. and Linda Trueb. *Biology of Amphibians.* Johns Hopkins University Press, 1994.

Although this textbook-style tome on all aspects of amphibian biology was published over thirty years ago, it remains a classic reference for the serious learner.

SCIENTIFIC LITERATURE

I have curated a list of the scientific studies that I consulted when writing this book. It is available on my website at EmilyTaylorScience.com.

ABOUT THE AUTHOR

Emily Taylor is a Professor of Biological Sciences at the California Polytechnic State University in San Luis Obispo, California, where she conducts research on the physiology, ecology, and conservation biology of reptiles and amphibians with her students. She got her bachelor's degree in English at UC Berkeley and her Ph.D. in Biology at Arizona State University. This is her third book, following *California Snakes and How to Find Them* (2024) and *California Lizards and How to Find Them* (2025). She lives in Atascadero with her husband, Steve, in their madhouse of rescued creatures, including Pax the dog, Aperol Spritz the bearded dragon, Baby the boa constrictor, Buzz and Snakeholio the rattlesnakes, and Helmut, Flash, and Bill the tortoises. Learn more at EmilyTaylorScience.com and follow her on social media @snakeymama.

Joe Johnston